CW01560483

Intermediate Problems

based on the
Intermediate Mathematical Challenge
1997–2016

Andrew Jobbings

The United Kingdom Mathematics Trust

Intermediate Problems

Published by The United Kingdom Mathematics Trust.

Maths Challenges Office, School of Mathematics, University of Leeds, Leeds LS2 9JT, United Kingdom

https://www.ukmt.org.uk

First published 2016

Reprinted 2017

ISBN 978-1-906001-29-2

Printed in the UK for the UKMT by The Charlesworth Press, Wakefield.

http://www.charlesworth.com

Typographic design by Andrew Jobbings of Arbelos.

http://www.arbelos.co.uk

Typeset with LaTeX.

The books published by the United Kingdom Mathematics Trust are grouped into series.

The EXCURSIONS IN MATHEMATICS series consists of monographs which focus on a particular topic of interest and investigate it in some detail, using a wide range of ideas and techniques. They are aimed at high school students, undergraduates and others who are prepared to pursue a subject in some depth, but do not require specialised knowledge.

1. *The Backbone of Pascal's Triangle*, Martin Griffiths

2. *A Prime Puzzle*, Martin Griffiths

The HANDBOOKS series is aimed particularly at students at secondary school who are interested in acquiring the knowledge and skills which are useful for tackling challenging problems, such as those posed in the competitions administered by the UKMT and similar organisations.

1. *Plane Euclidean Geometry: Theory and Problems*, A D Gardiner and C J Bradley

2. *Introduction to Inequalities*, C J Bradley

3. *A Mathematical Olympiad Primer*, Geoff C Smith

4. *Introduction to Number Theory*, C J Bradley

5. *A Problem Solver's Handbook*, Andrew Jobbings

6. *Introduction to Combinatorics*, Gerry Leversha and Dominic Rowland

7. *First Steps for Problem Solvers*, Mary Teresa Fyfe and Andrew Jobbings

8. *A Mathematical Olympiad Companion*, Geoff C Smith

The PATHWAYS series aims to provide classroom teaching material for use in secondary schools. Each title develops a subject in more depth and in more detail than is normally required by public examinations or national curricula.

1. *Crossing the Bridge*, Gerry Leversha

2. *The Geometry of the Triangle*, Gerry Leversha

The PROBLEMS series consists of collections of high-quality and original problems of Olympiad standard.

1. *New Problems in Euclidean Geometry*, David Monk

The CHALLENGES series is aimed at students at secondary school who are interested in tackling stimulating problems, such as those posed in the Mathematical Challenges administered by the UKMT and similar organisations.

1. *Ten Years of Mathematical Challenges: 1997 to 2006*

2. *Ten Further Years of Mathematical Challenges: 2006 to 2016*

3. *Intermediate Problems*, Andrew Jobbings

4. *Junior Problems*, Andrew Jobbings

❖

The YEARBOOKS series documents all the UKMT activities, including details of all the challenge papers and solutions, lists of high scorers, accounts of the IMO and Olympiad training camps, and other information about the Trust's work during each year.

Contents

II   More challenging problems

Appendix

Series Editor's Foreword

This book is part of a series whose aim is to help young mathematicians prepare for competitions at secondary school level. Here the focus is on the questions from the Intermediate Mathematical Challenge papers. Like other volumes in the Challenges series, it provides cheap and ready access to directly relevant material.

I hope that every secondary school will have these books in its library. The prices have been set so low that many good students will wish to purchase their own copies. Schools wishing to give out large numbers of copies of these books, perhaps as prizes, should note that discounts may be negotiated with the UKMT office.

London, UK   GERRY LEVERSHA

About the Author

Andrew Jobbings gained both his BSc and his PhD in mathematics from Durham University. He taught mathematics for 28 years, including 14 years as Head of Department at Bradford Grammar School, before founding the publishing business Arbelos.

With a keen interest in providing mathematics enrichment activities, Andrew devises problems for the UKMT and is involved with many other UKMT projects. He has regularly chaired a problems group for the European Kangaroo contest and gives Royal Institution masterclasses.

Preface

The Intermediate Mathematical Challenge (IMC) began in 1997, after the UKMT was founded. Prior to that, equivalent challenges were organised by Tony Gardiner, under the aegis of *The UK Mathematics Foundation*.

The problems on the IMC papers are intended to be stimulating as well as challenging—ideally, some problems will also raise a smile. That these aims are so admirably fulfilled is a measure of the quality of the problem setters.

Acknowledgements

Many people have been involved in the IMC since it began, including problem setters, checkers, and teachers in schools. Howard Groves has done sterling work over many years as Chair of the Problems Group. All of them, especially Tony Gardiner and Howard, deserve thanks for their help and support over the years of the competition.

I should particularly like to thank the following, who commented extremely helpfully on one or more draft versions of the book: Mary Teresa Fyfe; Howard Groves; Calum Kilgour; George and Stephen Power; Alan Slomson. There is no doubt that the book has improved immeasurably as a result.

Of course, any remaining mistakes are entirely the responsibility of the author.

Baildon, Shipley, UK   ANDREW JOBBINGS

Introduction

> I do believe that problems are the heart
> of mathematics ...
>
> ---
>
> Paul Halmos
> *The Heart of Mathematics*

Layout of the book

The Intermediate Mathematical Challenge (IMC) is a multiple-choice competition with 25 questions.

Problems

This book includes every problem used in the IMC from 1997 to 2016, but these are not given as multiple-choice questions. In most cases the five options have just been removed, but in a few cases, such as question 6 of exercise 1, the problem has been reworded to accommodate the options. In addition, the wording of some problems has been mildly edited.

The problems have been grouped together in two ways: by difficulty; and by topic.

Part I consists of problems appearing earlier on in the IMC papers (up to question 15); these are intended to be more straightforward than the later problems, which are given in part II.

In each of parts I and II, problems broadly based on the same topic have been grouped together into one exercise. In each exercise, the questions are ordered by their position on the original paper. The exercises themselves are roughly ordered by difficulty.

The allocation of problems to topic areas involves a great deal of subjectivity—and is sometimes a little arbitrary—so do not be surprised if you find a problem in an unusual place. Also, the names of the topics may bear little resemblance to anything you have already met; even when you have come across a topic before, the content could well be different to anything you may be used to. The real reason for grouping the problems, of course, is that it is convenient for the author to do so!

To help indicate the degree of difficulty of each question, the number of the question on the original paper is written in the left margin. This is a rough guide only, because the Problem Group's view of the difficulty of a question can be out of line with the outcome. In any case, removing the multiple-choice options may well affect the difficulty.

Remarks

Part III consists of remarks and answers.

Nearly every problem has a remark of some sort. In some cases the results needed for the given method are listed at the start of a remark, using the symbol '☛' to indicate each result.

The remarks are not intended to be 'full written solutions' (so would not get many marks in an Olympiad-style competition), but instead sketch out one method, providing a sequence of pointers so that, hopefully, by reading them you can solve a problem yourself. Sometimes a remark is just a collection of signposts and filling in the gaps may not be straightforward. Several of the remarks refer to non-standard methods.

Though it is not necessary if you are using this book, you may wish to read fuller solutions. These are provided by the UKMT in the solutions booklets, the Yearbooks, and (more recently) the extended solutions.

Note that the methods referred to in the remarks are not based on those given elsewhere: they may be the same; sometimes they are completely different.

Answers

The answer to every problem in the book is given in part III. The answers are upside down, to help you avoid reading them inadvertently.

Of course, the answer to a problem is unimportant in itself, other than as a (potential) check on the validity of the method used. What really matters is that you understand the underlying mathematics.

Using the book

The whole purpose of this book is to provide you with problems to solve. Removing the multiple-choice element means that you actually do need to solve the mathematical problem, rather than use other techniques (such as knowing that exactly one of the options is correct).

Select a problem and have a go. Use pencil and paper to do some calculations, or draw some diagrams, whatever is necessary to make a determined effort to solve the problem. In that way you will properly engage with the problem, and the mathematics contained in it. Do this *before* looking at part III. There are three possible outcomes.

✳ You do the problem and get it right.

   Well done! Even in this case it is worth reading any remarks: if your approach is the same as the one given, then you can confirm that your method is correct, otherwise, you may well learn something useful!

✳ You do the problem, but get it wrong.

   After checking that you have read the question correctly, see if knowing the correct answer enables you to find an error, either in your working or your approach. Otherwise, read any remarks and try again.

✳ You cannot do the problem.

   See if knowing the correct answer helps you to get there. Otherwise, read any remarks and try again. If you still have no joy, then ask someone.

You may find, partway through reading a remark, that you think 'Aha, gotcha, now I see how to do it!'. Then so much the better: stop reading and tackle the problem again. Conversely, you may find a remark too obscure. This is not deliberate (though the remarks are intended to leave you with some work to do). If you do find that you do not understand a remark and still cannot do a problem, try again later, or ask someone else.

As mentioned above, some remarks give a list of useful results. You may not yet have come across all of these; at the very least, try to find out why a result is true, either by proving it yourself, or perhaps by asking someone.

Often there is more than one way to tackle a problem, so your ideas may well differ from the method used in part III. This does not mean that your ideas are not valid—far from it—but it may be worth trying to remember the alternative approach because this could be useful in similar problems. Indeed, one measure of the quality of a problem is the number of different approaches that are possible.

Wait until you are ready before you tackle the harder problems; this applies particularly to the problems in part II.

When faced with unusual or challenging problems, what you need above all is perseverance, the desire to keep trying until some progress is made. Take your time when trying such a problem and keep puzzling away until it yields up its secrets. Success is rarely a question of extra knowledge, more often one of know-how.

In the IMC itself, of course, you are working against the clock. In such circumstances, having the relevant knowledge and know-how at your fingertips—being *fluent* in mathematics—is bound to be helpful. And that is where practice comes in, which brings us back to the purpose of this book.

Finally, suppose that you want to find a particular problem, one that you recall involves 'turnips', say. In that case you should try the index.

Calculators

Calculators are not allowed in the IMC.

In many of these problems a calculator offers no advantage, but many of the more arithmetical questions lose their point if you use a calculator.

You are advised not to use a calculator for any of the problems in the book.

Notation and terminology

You may occasionally find that the book uses notation or terminology which is not familiar to you, such as *km/h* for kilometres per hour, or *average* for what is sometimes called the mean.

Should you have any doubt about notation or terminology, please ask someone.

Part I

Problems

Arithmetic

Exercise 1

1. **1.** What is the value of $6102 - 2016$? *4086*

1. **2.** What is the value of $1 - 0.2 + 0.03 - 0.004$? *0.826*

1. **3.** What is the value of $4.5 \times 5.5 + 4.5 \times 4.5$? *45*

1. **4.** What is the value of $10 + 10 \times 10 \times (10 + 10)$? *2010*

1. **5.** What is the value of $1 + 2^3 + 4 \times 5$? *29*

1. **6.** When the following five numbers are arranged in increasing order of size, which one is in the middle?

    4.04   4.004   4.4   4.44   *4.044*

1. **7.** What is the value of $4002 - 2004$? *1998*

1. **8.** Between which of the following five pairs of numbers is there the greatest difference?

    −3, 8   −5, −13   1, 11   4, −5   −6, −15

2   **9.** To find the diameter in mm of a Japanese knitting needle, you multiply
     the size by 0.3 and add 2.1.   3.6
     What is the diameter in mm of a size 5 Japanese knitting needle?

2   **10.** Which of the following five calculations gives the greatest answer?

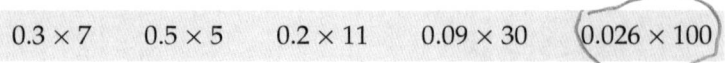

     0.3×7 0.5×5 0.2×11 0.09×30 0.026×100

3   **11.** Four of the following five numbers can make two pairs so that each
     pair adds up to 98 765.
     Which number is the odd one out?

     37 373   45 678   53 087   61 392   70 082

3   **12.** You are given that $2786 \times 231 = 643\,566$.   23106
     What is the value of $643\,566 \div 27.86$?

4   **13.** Ima Divvy used her calculator and multiplied a number by 20 instead
     of by 2.
     Which of the following could she do now to obtain the correct answer?

          divide by 20   divide by 40   multiply by 10
          multiply by 0.5   multiply by 0.1

4   **14.** A can of soup contains enough for three adults, or for five children.
     I have five cans of soup and fifteen children to feed.   6
     How many adults could I feed with what remains?

5   **15.** What is the value of $(12\,340 + 12.34) \div 1234$?   (0.01

5   **16.** Which of the following five numbers is equal to $30 \div 0.2$?

          1.5   6   15   150   600

Primes

Exercise 2

1 **1.** How many of the following four integers are prime?

$|$ 3   3$\cancel{3}$ $\cancel{338}$ $\cancel{3333}$

2 **2.** What is the smallest positive integer for which all the following are true?

    25

    (i) It is odd.

    (ii) It is not prime.

    (iii) The next largest odd integer is not prime.

2 **3.** What is the sum of the first five non-prime positive integers?   37

2 **4.** What is the largest prime that divides exactly into the number equal to $2 + 3 + 5 \times 7$?   5

3 **5.** How many of the following five expressions give answers which are *not* prime?

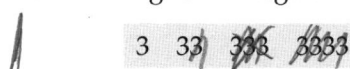

    0

3 **6.** Exactly one of the following five integers is prime.

2345   23 456   234 567   2 345 678   23 456 789

Which is it?

3 **7.** Which of the following is *not* prime?

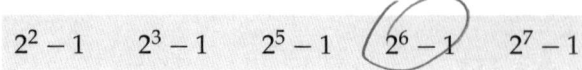

$2^2 - 1$ $2^3 - 1$ $2^5 - 1$ $2^6 - 1$ $2^7 - 1$

6 **8.** Observe that $2001 = 3 \times 23 \times 29$.

Which of the following five integers is also the product of exactly three distinct primes?

45   60   91   105   330

7 **9.** The positive integers p and q are the smallest primes that differ by 6. What is the sum of p and q?

Fractions

Exercise 3

1   **1.** What fraction is equal to 25% of $\frac{3}{4}$? $\frac{9}{16}$

1   **2.** What fraction has a value that is halfway between $\frac{1}{4}$ and $\frac{1}{6}$? $\frac{5}{24}$

1   **3.** What is the value of 3 divided by $\frac{1}{2}$? 6

1   **4.** Which of the following five numbers could replace $\heartsuit$ so that the value of $\frac{\heartsuit}{5}$ lies between 3 and 4?

$$3.2 \qquad 9 \qquad 14 \qquad 19 \qquad 24$$

1   **5.** One quarter of a number is 24.
What is one third of the original number? 32

2   **6.** Which of the following five fractions has a value that is closest to 1?

$$\frac{7}{8} \qquad \frac{8}{7} \qquad \frac{9}{10} \qquad \frac{10}{11} \qquad \frac{11}{10}$$

3   **7.** What is the value of a half of a third, plus a third of a quarter, plus a quarter of a fifth?

$$\frac{18}{60} \quad or \quad \frac{3}{10}$$

3   **8.** What is the value of $0.75 \div \frac{3}{4}$?

3   **9.** What fraction has a value that is halfway between $\frac{1}{4}$ and $\frac{1}{8}$? $\overset{3}{)}$

3   **10.** Which of the following five fractions has the largest value? $1\,6$

$$\frac{7}{15} \qquad \frac{3}{7} \qquad \frac{11}{23} \qquad \frac{4}{9} \qquad \frac{6}{11}$$

4   **11.** Between them, Ginger and Victoria eat two thirds of a cake. Ginger
eats one quarter of the cake.

What fraction of the cake does Victoria eat? $\frac{5}{12}$

4   **12.** Which of the following five fractions is not equal to the others?

$$\frac{1+4}{7+4} \qquad \frac{20}{140} \qquad \frac{0.2}{1.4} \qquad \frac{1 \times 11}{7 \times 11} \qquad \frac{8}{56}$$

5   **13.** Which of the following five expressions has the greatest value?

one half of $\dfrac{1}{25}$ one third of $\dfrac{1}{20}$ one quarter of $\dfrac{1}{15}$

one fifth of $\dfrac{1}{10}$ one sixth of $\dfrac{1}{5}$

5   **14.** Which of the following five fractions is *not* equal to an integer?

$$\frac{594}{5+9+4} \qquad \frac{684}{6+8+4} \qquad \frac{756}{7+5+6} \qquad \frac{873}{8+7+3} \qquad \frac{972}{9+7+2}$$

5   **15.** What is the value of $\dfrac{67 \times (67 + 67)}{67}$? 134

6   **16.** One third of the animals in Jacob's flock are goats, the rest are sheep.
There are twelve more sheep than goats.

How many animals are there altogether in Jacob's flock?

36

6 **17.** What fraction has a value that is halfway between $\frac{4}{5}$ and $\frac{2}{3}$?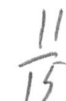

7 **18.** Which of the following five fractions has the smallest value?

$$\frac{2+3}{4+6} \qquad \frac{2 \div 3}{4 \div 6} \qquad \frac{23}{46} \qquad \frac{2-3}{4-6} \qquad \frac{2 \times 3}{4 \times 6}$$

Integers

Exercise 4

1   **1.** Which of the following is divisible by 6?

> one million minus one   one million minus two
> one million minus three   one million minus four
> one million minus five

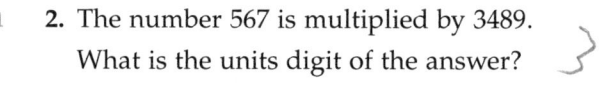

1   **2.** The number 567 is multiplied by 3489.
What is the units digit of the answer? 3 ✓

2   **3.** Three positive integers are all different. Their sum is 7.
What is their product? 8

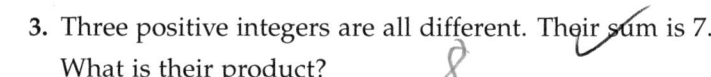

2   **4.** What is

> ten million one hundred thousand and one

when written in digits rather than words?

10,100,001 ✓

3   **5.** The consecutive digits 1, 2, 3, 4 in that order can be arranged to make
     the correct division

$$12 \div 3 = 4.$$

One other sequence of four consecutive digits makes a correct division

$$'pq' \div r = s.$$

What is the value of *s* in this case?   8   ✓

3   **6.** Alice's room is furnished with three-legged stools and four-legged
     chairs. There are 17 legs in all—excluding Alice's!
     How many three-legged stools are there?   3   ✓

4   **7.** Amrita is baking a cake today (Thursday). She bakes a cake every fifth
     day.
     How many days will it be before she next bakes a cake on a Thursday?
        35   ✓

4   **8.** Suppose that *m* is an even integer and *n* is an odd integer.
     Which of the following is an odd integer?   ✓

$$3m + 4n \qquad 5mn \qquad (m + 3n)^2 \qquad m^3 n^3 \qquad 5m + 6n$$ ✓

4   **9.** All four digits of two positive two-digit integers are different.
     What is the largest possible sum of two such integers?   183   ✓

4   **10.** Old Martha has 5 children, each of whom has 4 children, each of
      whom has 3 children, each of whom is childless.
      How many descendants does Old Martha have?   60

5   **11.** My local greengrocery *'Apples and Pears'* charges 24p for the first apple
      bought, 23p for the second, 22p for the third, and so on; each apple
      costs 1p less than the one before.
      How much change should I receive when I buy 9 apples and I give
      the shopkeeper £2?   20p   ✓

6 **12.** When I glanced at my car milometer it showed 24942, a palindromic number. Two days later, I noticed that it showed the next palindromic number.

How many miles did my car travel in those two days?   *110* ✓

6 **13.** The sum of nine consecutive positive integers is 2007.

What is the difference between the largest and smallest of these integers?   *8*   ✓

6 **14.** Think of an integer, double it, then add 3. Multiply your answer by 4 and from this take away 5. Now also take away the number you first thought of.

No matter what your first number was, your answer will always be a multiple of one particular integer.

Which of the following is that integer?   ✓

2   3   5   (7)   11

6 **15.** Granny has been having a smashing time. Yesterday she had 12 cups and 10 matching saucers, but this morning she dropped a tray holding one third of the cups and half the saucers, breaking all of those on the tray.

How many of her cups are now without saucers?

3   ✓

Time and dates

Exercise 5

1. **1.** How many hours are there in a week in which the clocks do not change?

1. **2.** At midnight on 15 December 2005, the moon reached its highest point in the sky, an event that occurs every 18.6 years.

   In what year after that will it next occur?

2. **3.** A machine cracks open 180 000 eggs per hour.

   How many eggs is that per second?

3. **4.** A certain company offers "750 hours of free internet use for new subscribers". On closer inspection it becomes clear that this time must be used during the new subscriber's first month of membership!

   What is the maximum number of hours in any one month of the year (ignore the fact that in some months the clocks change)?

4. **5.** A radio advertisement claimed:

   > Use ———— artificial sweetener every day
   > and save 7000 calories in a year.

   To the nearest ten, how many calories is this per day?

5   **6.** A ship's bell is struck every half-hour, starting with one bell at 00:30, two bells (meaning the bell is struck twice) at 01:00, three bells at 01:30 until the cycle is complete with eight bells at 04:00. The cycle then starts again with one bell at 04:30, two bells at 05:00, and so on.

What is the total number of times the bell is struck between 00:15 on one day and 00:15 on the following day?

5   **7.** How many minutes will elapse between 20:12 today and 21:02 tomorrow?

6   **8.** In Britain in 1996 we consumed on average 9.6 kg of bananas per person per year (that is, around 60 bananas each). In some parts of Africa, the consumption of bananas is as high as 250 kg of bananas per person per year.

Which of the following roughly corresponds to that consumption of bananas?

4 or 5 a day	1 or 2 a day	4 or 5 a week	1 or 2 a week
4 or 5 a month			

7   **9.** The Three Choirs Festival is held annually. Its venue rotates in a three-year cycle among Hereford, Gloucester and Worcester.

In 2009 it was held in Hereford, in 2010 it was held in Gloucester, the next year it was held in Worcester.

Assuming that this three-year cycle continues, in which one of the following five years will the Festival *not* be held in Worcester?

2020	2032	2047	2054	2077

8   **10.** On my clock's display, the time has just changed to 02:31.

How many minutes will it be until all the digits 0, 1, 2, 3 next appear together again?

8   **11.** In a certain year, there were exactly four Tuesdays and exactly four Fridays in October.

On what day of the week did Halloween, 31 October, fall that year?

8 **12.** Sydney flew to Melbourne, Australia. The flying time to Melbourne, which is 11 hours ahead of Britain, was 21 hours. Sydney's flight left London at 11.30 am on Tuesday.

What time was it in Melbourne when Sydney's flight arrived?

8 **13.** School starts at 8:30 am and finishes at 3:15 pm.

What fraction of a 24-hour day does school take up?

8 **14.** In March 2003 Welshman Tony Evans dropped a ball from an aircraft a mile above the Mojave desert to see if it would bounce. The ball was made from six million rubber bands, had a circumference of 14 ft 8 in, weighed 2600 pounds and took Mr Evans five years to build.

Which of the following is, on average, roughly the number of rubber bands that he added each day whilst building the ball?

$$3 \qquad 33 \qquad 330 \qquad 3300 \qquad 33\,000$$

9 **15.** Which of the following is the best estimate for the number of seconds between the start of the year 2000 and 1 February 2001?

$$3 \times 10^4 \qquad 3 \times 10^5 \qquad 3 \times 10^6 \qquad 3 \times 10^7 \qquad 3 \times 10^8$$

9 **16.** Leap years normally occur every four years. However, years at the turn of a century are leap years only when they are multiples of 400. Therefore the year 2000 was a leap year, but the year 1900 was not a leap year.

How many leap years will there be between 2001 and 3001?

10 **17.** One year in the 1990s, 1 January fell on a Monday. Eleven years later, 1 January was also a Monday.

How many times did 29 February occur during those eleven years?

10 **18.** The time shown on a digital clock is 5:55.

How many minutes will pass before the clock next shows a time for which all the digits are the same?

10 **19.** In March 1998 a book called *The Shadow of the East* was returned to a
    library in Sussex. It had been borrowed on 3 January 1924! The library
    charges a fine of 60p per week for overdue books.

    Which of the following is approximately the fine the person who
    returned the book should have paid?

    £45   £180   £230   £2200   £16 000   ✓

11 **20.** The *World Wide Fund for Nature* estimates that 54 acres of Brazilian
    rainforest are destroyed every minute of every day.

    Which of the following is roughly the number of acres lost each week?

    50 000   80 000   200 000   500 000   2 000 000

    ✓

Miscellany 1

Exercise 6

2   **1.** An ice-cream stall sells vanilla, strawberry and chocolate ice-creams. The pie chart illustrates the sales of ice-cream for last Saturday. The number of vanilla and the number of chocolate ice-creams sold were the same. The stall sold 60 strawberry ice-creams.

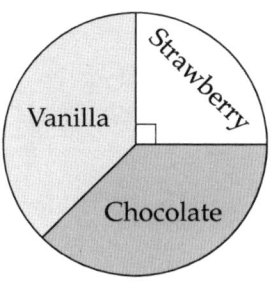

How many chocolate ice-creams were sold?

2   **2.** A sheet of A4 paper (297 mm × 210 mm) is folded once and then laid flat on a table.

Which of the following five shapes could not be made?

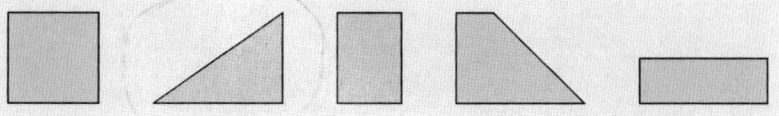

3   **3.** In how many ways can a square be cut in half using a single straight line cut?

3   **4.** A fair die has just been rolled five times, giving scores of 1, 2, 3, 4, 5 in that order.

What is the probability that the score on the next roll is a 6?

4   **5.** The diagram shows an equilateral triangle divided into small congruent equilateral triangles.

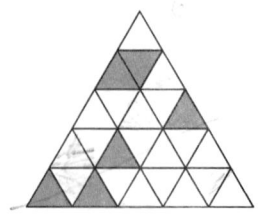

What is the lowest number of small triangles which have to be shaded (in addition to those shown) to produce a figure which has a line of symmetry?

4   **6.** In Worcestershire, Wyre Piddle is 12 km south of the village of North Piddle and Abbotts Morton is 12 km east of North Piddle.

What is the direction of Abbotts Morton from Wyre Piddle?

5   **7.** When travelling from London to Edinburgh by train, you pass a sign saying

Edinburgh 200 miles

Then, $3\frac{1}{2}$ miles later, you pass another sign saying

Halfway between London and Edinburgh

How many miles is it by train from London to Edinburgh?

6   **8.** In Niatirb they use Cibara numerals. These are the same shape as normal Arabic numerals, but with the meanings in the opposite order. So "0" means "nine", "1" means "eight" and so on. However, they write their numbers from left to right and use arithmetic symbols just as we do. So, for example, they use 62 for the number we write as 37.

How do the inhabitants of Niatirb write the answer to the sum which they write as $837 + 742$?

6 **9.** The 'letter-product' of an integer is obtained by multiplying the integer by the number of letters in the corresponding word. For example, the letter-product of 5 is 20, since there are 4 letters in the word 'five' and $5 \times 4 = 20$.

Which of the following five integers has the largest letter-product?

6   7   8   9   10

6 **10.** *PQRS* is a square with sides of length 9 cm.

How many points (inside or outside the square) are equidistant from *Q* and *R*, and are exactly 6 cm from *P*?

8 **11.** How large will an angle of 2.5° appear to be when you enlarge it by looking through a stack of five magnifying glasses, each one of which magnifies by a factor of two?

9 **12.** Jane has 20 identical cards in the shape of a right-angled isosceles triangle. She uses the cards to make the five shapes below.

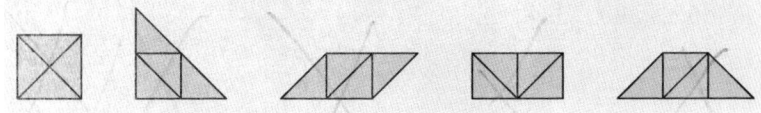

Which of the shapes has the shortest perimeter?

9 **13.** The diagram shows 10 identical coins which fit exactly inside a wooden frame. As a result each coin is prevented from sliding.

What is the largest number of coins that may be removed so that each remaining coin is still unable to slide?

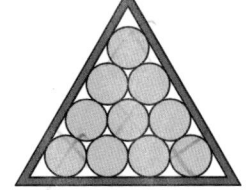

9 **14.** It is evening and Meg, who is 1 m tall, casts a shadow of length 3 m. Meg stands on her brother's shoulders, which are 1.5 m above the ground.

How long a shadow will she and her brother cast?

11 **15.** The diagram below shows a pattern which repeats every 12 dots.

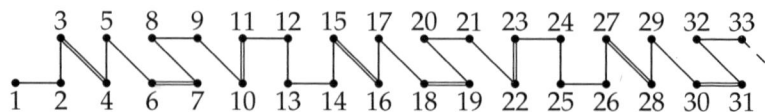

The section between 2007 and 2011 look like looks like one of the following. Which one?

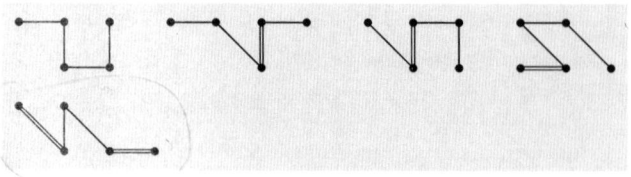

12 **16.** The sum

$$one + four = seventy$$

becomes correct when each word is replaced by the number of letters in it to give

$$3 + 4 = 7.$$

Using the same convention, which of the following five words could be substituted for x to make the sum

$$three + five = x$$

true?

    eight   nine   twelve   seventeen   eighteen

13 **17.** The diagram shows that

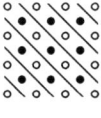

$$1 + 3 + 5 + 7 + 5 + 3 + 1 = 3^2 + 4^2.$$

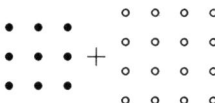

Which of the following five expressions is equal to

$$1 + 3 + 5 + \cdots + 1999 + 2001 + 1999 + \cdots + 5 + 3 + 1?$$

$999^2 + 1000^2$	$1000^2 + 1001^2$	$1999^2 + 2000^2$
$2000^2 + 2001^2$	none of these	

13 **18.** I fold a piece of paper in half, then in half again before cutting (and discarding) a shape from the folded paper as shown.

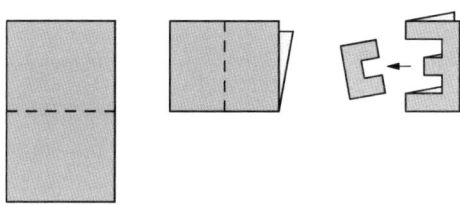

When I unfold the paper, what do I see?

15 **19.** A designer wishes to use two copies of the pattern of dots shown on the right to create a new pattern, without any of the dots overlapping.

Which one of the following five patterns could be made?

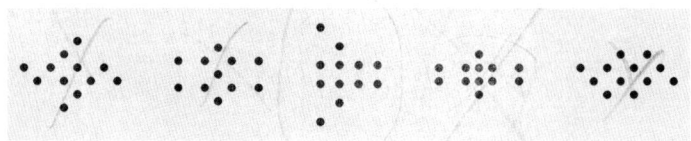

15   **20.** Jack had five cards in a pile on a table. He gave me the top card, and then placed the next card at the bottom of the pile; then he gave me the next one on the top and placed the next one after that at the bottom of his pile. He continued like this until he had given me all of the cards. I looked down and to my surprise found that Jack had given me the cards in order: Ace, 2, 3, 4, 5.

In what order (top to bottom) did Jack originally have the cards in the pile on the table?

Powers and square roots

Exercise 7

1. Which of the following has the greatest value?

$$19 \times 99 \qquad 199 \times 9 \qquad 199^9 \qquad 1^9 \times 9^9 \qquad 1^{999}$$

2. Which of the following five integers is not a square?

$$16 \qquad 36 \qquad 64 \qquad 80 \qquad 100$$

3. Which of the following five expressions is equal to 2005?

$$(1^2 + 1)(10^2 + 1) \qquad (2^2 + 1)(20^2 + 1) \qquad (3^2 + 1)(30^2 + 1)$$
$$(4^2 + 1)(40^2 + 1) \qquad (5^2 + 1)(50^2 + 1)$$

4. Which of the following five integers is not a square?

$$1^6 \qquad 2^5 \qquad 3^4 \qquad 4^3 \qquad 5^2$$

5. What is the value of 2003^2?

6   **6.** Which of the following five expressions has the least value?

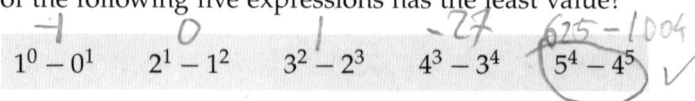

$$1^0 - 0^1 \qquad 2^1 - 1^2 \qquad 3^2 - 2^3 \qquad 4^3 - 3^4 \qquad 5^4 - 4^5$$

8   **7.** How many two-digit squares differ by 1 from a multiple of 10?

8   **8.** What is the value of $2^{10} - 10^2$?

13   **9.** Which of the following five expressions is equal to half of 2^{20}?

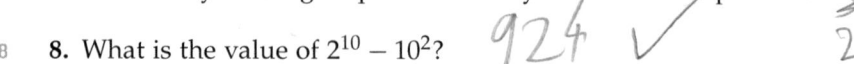

$$1^{10} \qquad 1^{20} \qquad 20 \qquad 2^{10} \qquad 2^{19}$$

14   **10.** The number 6 lies exactly halfway between 3 and 3^2.

Which of the following five numbers is *not* halfway between a positive integer and its square?

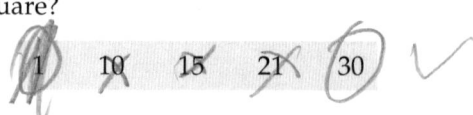

$$1 \qquad 10 \qquad 15 \qquad 21 \qquad 30$$

14   **11.** Catherine's computer correctly calculates $\dfrac{66^{66}}{2}$.

What is the units digit of its answer?

15   **12.** How many of the five numbers

$$3\sqrt{11} \qquad 4\sqrt{7} \qquad 5\sqrt{5} \qquad 6\sqrt{3} \qquad 7\sqrt{2}$$

are greater than 10?

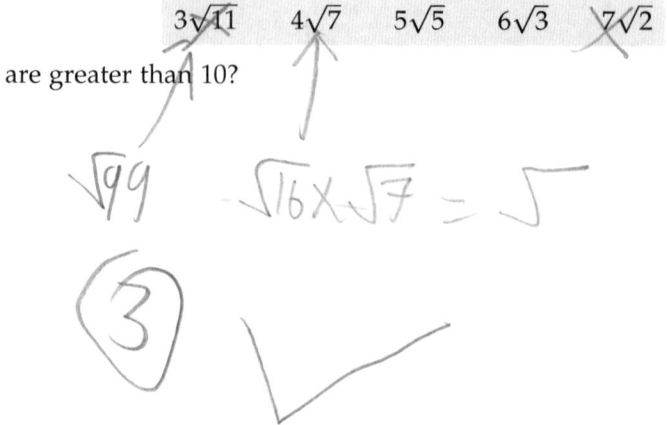

(handwritten working at top)

64×16
$\frac{}{}$
364
64
1004

$\cancel{9} 16, 25, 36, 49, 64, 81$

$2 \times 2 \times 2 \times 2 \times 2 = 2^5 = 32$
$4 \times 4 \times 2$

32×32 ... $9 \, 64$

64
$16 \, 48 \, 19 \, 6 \, 0$
$1 \, 8 \, 2 \, 4$

Reasoning

Exercise 8

2   **1.** The diagram shows seven metal rings linked together.

What is the smallest number of rings that need to be cut in order to separate all the rings?   *4 3*

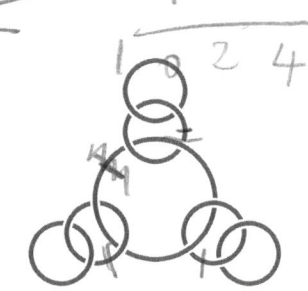

2   **2.** You are told that 30 pupils have 25 different birthdays between them.

What is the largest number of these pupils who could share the same birthday?   *6 ✓*

2   **3.** Four of the following five tiles may be put side by side so that they simultaneously spell two imperial units of length.

Which tile is left out?

In Mi
Ft NM
Yd

E	I	L	K	M
D	A	R	C	Y

30
I	A
M	
Y	
D	E

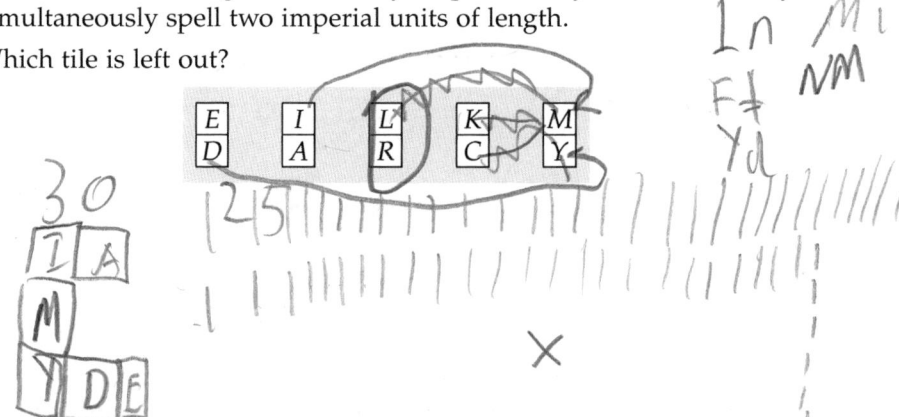

5   **4.** How many of the statements in the box are true?

> Any number that is divisible by 6 is even.
> Any number that is divisible by 9 is odd.
> The sum of any two odd numbers is even.
> The sum of any two even numbers is odd.

5   **5.** In a magic square, each row, each column and both main diagonals have the same total.

In the partially completed magic square shown, what number should replace N?

18			
13	15		
20	10	11	17
7	N	16	14

5   **6.** J is the set of High Court judges;
K is the set of living things whose names begin with K;
L is the set of all living creatures;
M is the set of brilliant mathematicians.

Kevin is a very ordinary kangaroo.

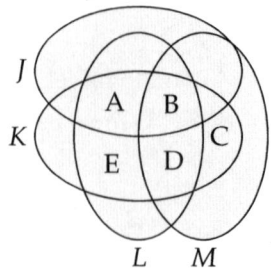

In which of the five regions A, B, C, D and E of the Venn diagram does Kevin belong?

7   **7.** The Queen of Spades always lies for the whole day or always tells the truth for the whole day.

Which of the following five statements can she never say?

> Yesterday, I told the truth.   Yesterday, I lied.
> Today, I tell the truth.   Today, I lie.
> Tomorrow, I shall tell the truth.

8   **8.** Seb has been challenged to place the digits from 1 to 9 in the nine
regions formed by the Olympic rings so that there is exactly one digit
in each region and the sum of the digits in each ring is 11. The diagram
shows part of his solution.

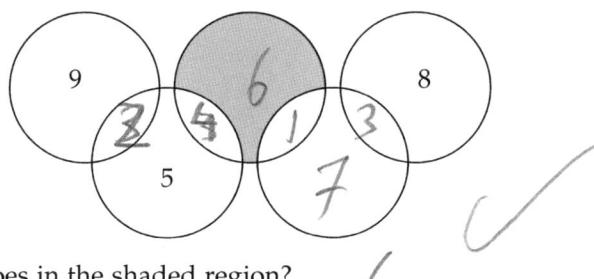

What digit goes in the shaded region? 6

9   **9.** Which of the following five numbers is the total number of letters in
the other four? 9

eleven   twenty-two   thirty-three   forty-four   fifty-five

11   **10.** Not all characters in the *Woodentops* series tell the truth. When Mr
Plod asked them, "How many people are there in the Woodentops
family?", four of them replied as follows:

> Jenny: "An even number."
> Willie: "An odd number."
> Sam: "A prime number."
> Mrs Scrubitt: "A number which is the product of two integers
> greater than one."

How many of these four were telling the truth? 2

11   **11.**   "You eat more than I do," said Tweedledee to Tweedledum.
"That is not true," said Tweedledum to Tweedledee.
"You are both wrong," said Alice, to them both.
"You are right," said the White Rabbit to Alice.

How many of the four statements were true? 4

12 **12.** In the game Four-in-a-Row, two players take it in turns to place counters on the 5 × 5 board shown. The winner is the first player to have four adjacent counters in a line across or down (but not diagonally).

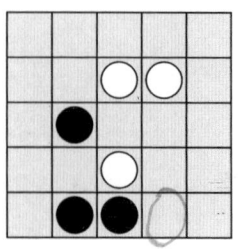

In the diagram, it is Black's turn to play next.

Where should Black play her fourth counter to be certain of winning on her fifth turn whatever White plays?

13 **13.** The diagram shows a heptagon with a line of three circles on each side. Each circle is to contain exactly one number. The integers 8 to 14 are distributed as shown and the integers 1 to 7 are to be distributed to the remaining circles.

The total of the integers in each of the lines of three circles is to be the same.

What is this total?

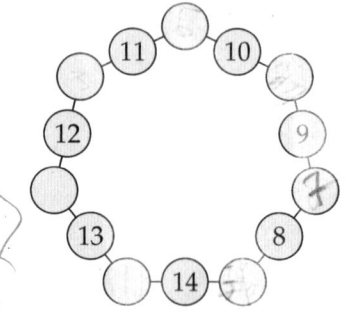

13 **14.** Granny has made another of her special super-heavy giant rock cakes. At her birthday party, five of the guests tried to guess the weight of the cake. Their guesses were 5040 g, 5060 g, 5110 g, 5120 g and 5150 g. Actually, none of these guesses was right. Only two were more than 30 grams out, and they were out by 70 g and 90 g.

What was the weight of the cake? *5130*

14 **15.** Tegwen has the same number of brothers as she has sisters. Each one of her brothers has 50% more sisters than brothers.

How many children are in Tegwen's family?

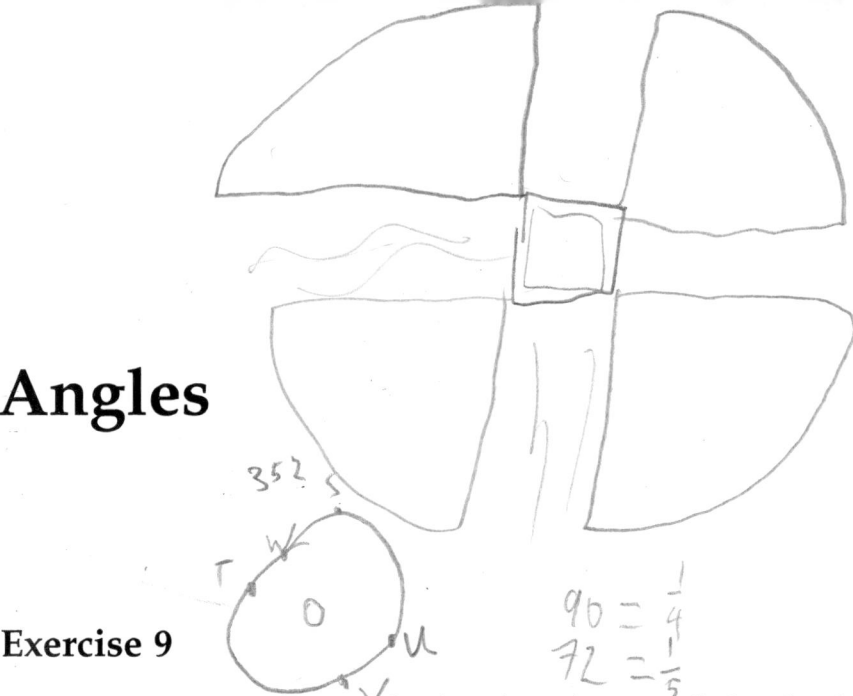

Angles

Exercise 9

3   **1.** Starting at a point S on a fixed circle with centre O, I make the following moves, in order:

    (i) travel anticlockwise one quarter of the way round the circle to a point T;

    (ii) hop across to U—the point at the opposite end of the diameter through T;

    (iii) travel one fifth of the way round the circle clockwise to the point V;

    (iv) hop across to W—the point at the opposite end of the diameter through V.

What is the size of angle WOS?

4   **2.** The angles of a triangle are in the ratio $2 : 3 : 5$.

What is the difference between the largest angle and the smallest angle?

4   **3.** The diagram shows two isosceles triangles, in which the four angles
     marked $x°$ are equal. The two angles marked $y°$ are also equal.

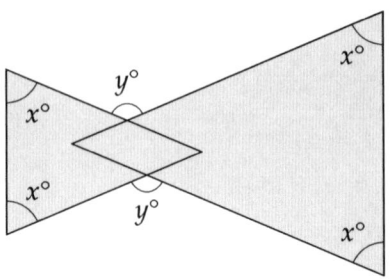

Which of the following five statements is always true?

$$y = 2x \qquad y = x + 30 \qquad y = x + 60 \qquad y = x + 90$$
$$y = 180 - x$$

4   **4.** What is the value of

$$a + b + c + d + e + f?$$

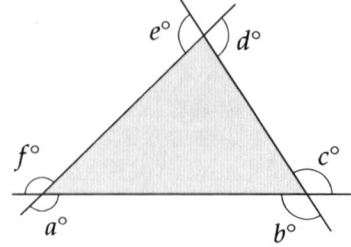

5   **5.** In the diagram, the lines
     PQ and SR are parallel,
     as are the lines PS and
     QT.

     What is the value of x?

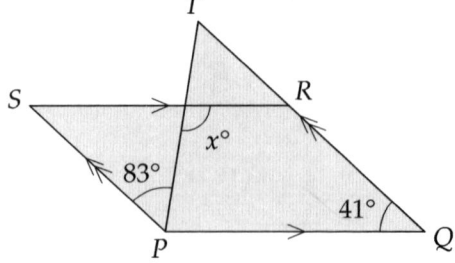

6   **6.** In triangle *PQR*, the point *S* lies on *QR* so that *QS* = *SP* = *PR* and
     ∠*SPQ* = 20°.

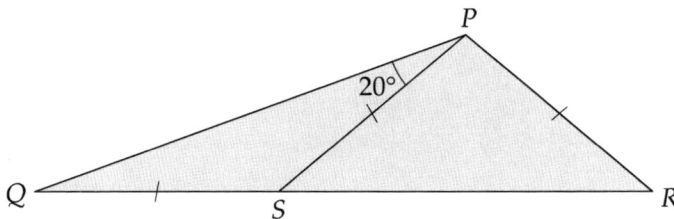

    What is the size of ∠*SRP*?

7   **7.** In the diagram, what is the value of *x*?

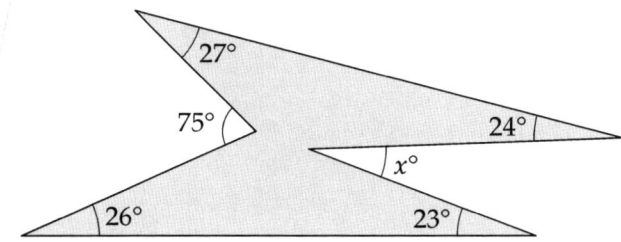

7   **8.** What is the value of *x* in this
     diagram?

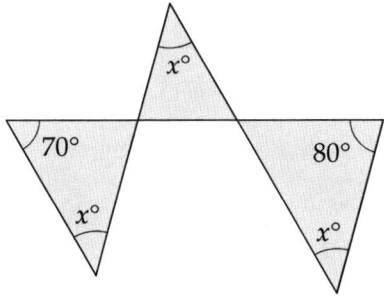

8   **9.** In the diagram, lines *PQ* and *RS* are
     parallel and *QR* = *QS*. The marked
     angle is acute but not equal to 60°.

     How many other angles in the
     diagram are equal to the marked
     angle?

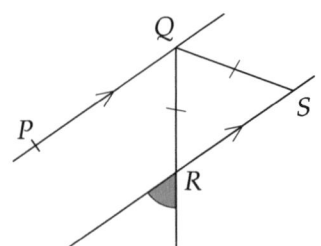

8   **10.** In the diagram, *RP* = *PS* =
     SQ and *QSR* is a straight
     line.

     What is the value of *x*?

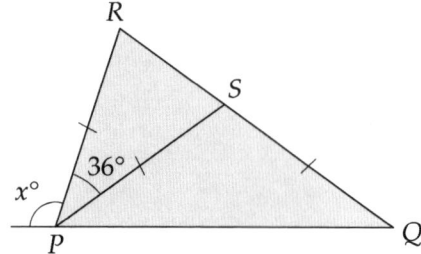

9   **11.** In the diagram, *XY* is a straight line.

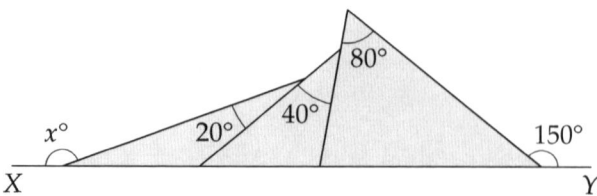

     What is the value of *x*?

10   **12.** The angles of a quadrilateral taken in order are $x°$, $5x°$, $2x°$ and $4x°$.

     Which one of the following five terms correctly describes the quadri-
     lateral?

     ~~kite~~   ~~parallelogram~~   rhombus   ~~arrowhead~~   trapezium

11 **13.** The diagram shows a parallelogram *PQRS* in which ∠*QRS* = 50°.
The point *T* is such that ∠*TPS* = 62° and *TP* = *PQ*.

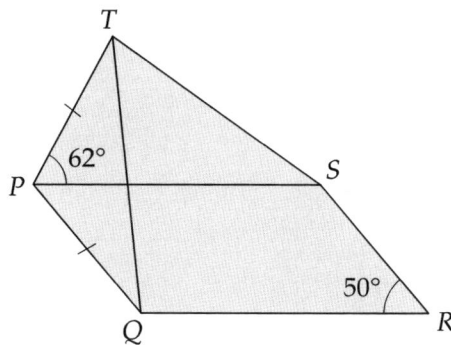

What is the size of ∠*TQR*?

11 **14.** What is the obtuse angle between the hands of a clock at 6 minutes
past 8 o'clock?

11 **15.** In the quadrilateral *PQRS*,
∠*PQR* = 90°, ∠*SPQ* = 70° and
QR = *QS* = *QP*.
What is the size of ∠*RSQ*?

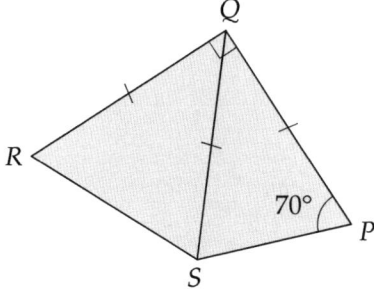

13 **16.** The diagram shows two equilateral
triangles.
What is the value of *x*?

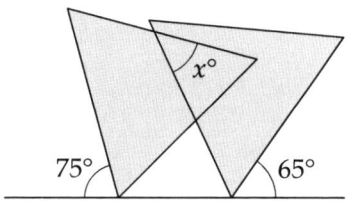

14   **17.** Ten stones, of identical shape and
size, are used to make an arch, as
shown in the diagram. Each stone
has a cross-section in the shape of
a trapezium with three equal sides.

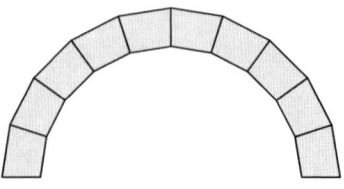

What is the size of the smallest
angles of the trapezium?

More arithmetic

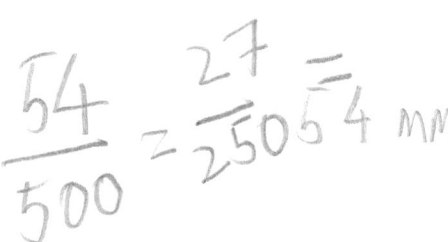

Exercise 10

7   1. A ream of paper (500 sheets) is 5·4 cm thick.

How many millimetres thick is a single sheet, correct to one significant figure?

7   2. The answers to the following five calculations are placed in order.

Which calculation gives the answer in the middle?

$$19 + 97 \qquad 19 - 97 \qquad 19 \times 97 \qquad 19 \div 97 \qquad 19^{97}$$

8   3. What is the value of $2.017 \times 2016 - 10.16 \times 201.7$?

9   4. It has been estimated that the mass of insects caught by spiders in a year in the UK is equal to the mass of the human population of the UK.

Assuming this population is 60 million and the average mass of a human is 70 kg, how many tonnes of insects are caught by spiders per year in the UK?

10   5. Merlin magically transforms a 6 tonne monster into mice with the same total mass. Each mouse has a mass of 20 g.

How many mice does Merlin make?

10   **6.** Gill is 18 this year. She and I went to a restaurant for lunch to celebrate her birthday lunch.

The bill for lunch for the two of us came to £25.50. Gill paid the bill by credit card and I left a £2.50 tip in cash. We agreed to split the total cost equally.

How much did I owe Gill?

10   **7.** Mickey Mouse wants to buy 20 g of cheese. Cheese costs £3.41 per kilo.

Which of the following five calculations would Mickey need to do to find the cost, in pence, of 20 g of cheese?

3.41×20 $3.41 \div 20$ 3.41×0.02 $341 \div 0.02$ 341×0.02

Polygons

Exercise 11

2   **1.** Three of the interior angles of a given quadrilateral are each 80°.
      What is the fourth angle of this quadrilateral?

3   **2.** An equilateral triangle is placed inside
      a larger equilateral triangle, as shown,
      so that the diagram has three lines of
      symmetry.

      What is the value of x?

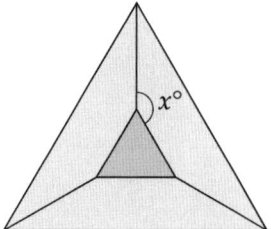

3   **3.** An equilateral triangle, a square and a regular pentagon
      all have the same side-length. The triangle is drawn on
      and above the top side of the square and the pentagon
      is drawn on and below the bottom side of the square,
      as shown.

      What is the sum of the interior angles of the resulting
      polygon?

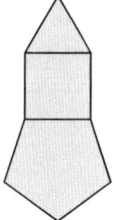

4   **4.** The diagram shows a regular pentagon inside a
       square.

       What is the value of x?

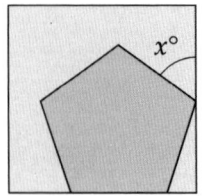

7   **5.** In the diagram, what is the sum of
       the marked angles?

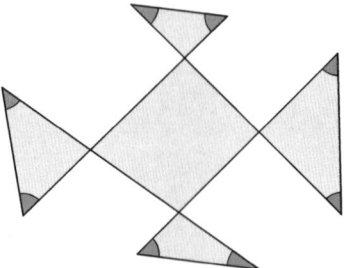

9   **6.** What is the value of

       $$p + q + r + s + t + u + v + w + x + y$$

       in the diagram?

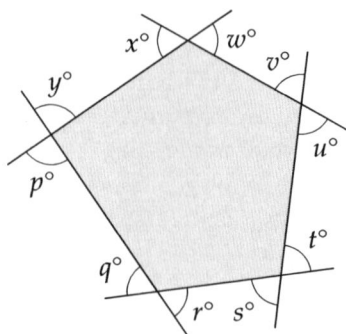

10   **7.** *PQRST* is a regular pentagon and *QRU* is an equilateral triangle such
       that *U* is inside *PQRST*.

       What is the size of $\angle UPQ$?

10   **8.** The diagram shows three squares drawn
      on the sides of a triangle.

      What is the sum of the three marked
      angles?

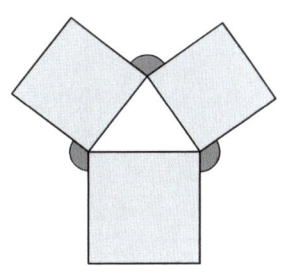

10   **9.** Equal regular pentagons are placed together
      to form a ring in the manner shown. The
      diagram shows the first three pentagons.

      How many more pentagons are needed to
      complete the ring?

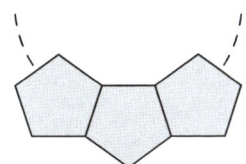

12   **10.** The diagram shows a square inside a regular
      hexagon.

      What is the value of x?

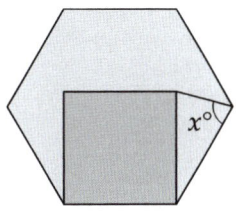

14   **11.** One of the following five statements is false. Which one?

      an octagon has twenty diagonals
      a hexagon has nine diagonals
      a hexagon has four more diagonals than a pentagon
      a pentagon has the same number of diagonals as it has sides
      a quadrilateral has twice as many diagonals as it has sides

15   **12.** The diagram shows a regular pentagon
      PQRST. The lines *QS* and *RT* meet at *U*.

      What is the size of angle *RUP*?

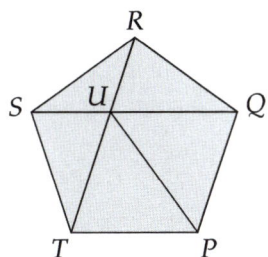

15   **13.** Each interior angle of a particular polygon is an obtuse angle which is a whole number of degrees.

What is the greatest number of sides the polygon could have?

Percentages

Exercise 12

2   **1.** What is the difference in value between 10% of one million and 10% of one thousand?

2   **2.** What is the value of 6% of 6 plus 8% of 8?

3   **3.** Which of the following five calculations gives the greatest value?

       50% of 10   40% of 20   30% of 30   20% of 40   10% of 50

4   **4.** A standard pack of pumpkin seeds contains 40 seeds. A special pack contains 25% more seeds. Rachel bought a special pack and 70% of the seeds germinated.

     How many pumpkin plants did Rachel have?

4   **5.** What percentage of the large 5 × 5 square is shaded?

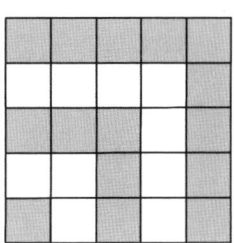

8 **6.** GOOD NEWS! Accidents in British homes involving tea-cosies went down from three in 1993 to zero in 1994.

What percentage decrease was that?

9 **7.** Auntie Fi's dog Itchy has a million fleas. His anti-flea shampoo claims to leave no more than 1% of the original number of fleas after use.

What is the least number of fleas that will be eradicated by the treatment?

9 **8.** Joseph's flock has 55% more sheep than goats.

What is the ratio of goats to sheep in the flock?

11 **9.**

> S is 25% of 60.
>
> 60 is 80% of U.
>
> 80 is M% of 25.

What is the value of $S + U + M$?

11 **10.** My bargain settee cost me £240 in a sale offering 25% reductions on all items.

How much did I save?

11 **11.**

> 20% off everything

screamed the sale posters. I paid £60.

What would I have paid before the sale?

12 **12.** Which one of the following five calculations gives a different value from all the others?

> 18% of £30   12% of £50   6% of £90   4% of £135
> 2% of £270

13 **13.** A 30 cm × 40 cm page of a book includes a 2 cm margin on all four edges, as shown.

What percentage of the page is occupied by the margins?

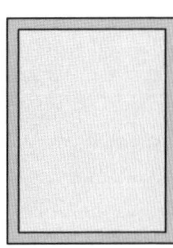

13 **14.** What is the value of 50% of 2006 plus 2006% of 50?

14 **15.** The Head triumphantly announced

> "80% of the pupils asked preferred purple blazers, so purple blazers it is!"

conveniently forgetting to mention that 80% of pupils of the school had not even been asked.

What percentage of pupils had actually told the Head that they preferred purple?

15 **16.** To make porridge, Goldilocks mixes together 3 bags of oats with 1 bag containing 20% wheat bran and 80% oats. All the bags have the same volume.

What percentage of the volume of Goldilocks' porridge mixture is wheat bran?

Rates

Exercise 13

2   **1.** Last year, Australian Suzy Walsham won the annual women's race up the 1576 steps of the Empire State Building in New York for a record fifth time. Her winning time was 11 minutes 57 seconds.

Which of the following is roughly the number of steps she climbed per minute?

13	20	80	100	130

4   **2.** The coach of the Irish hockey team has a maximum speed of 60 miles per hour. The coach travels at this speed for two hours.

Which of the following is roughly the number of *kilometres* that the coach travels?

120	160	200	240	280

5   **3.** The northern wheatear is a small bird weighing less than an ounce. Some northern wheatears migrate from sub-Saharan Africa to their Arctic breeding grounds, travelling almost 15 000 km. The journey takes just over 7 weeks.

Which of the following is roughly the average distance they travel each day?

1 km   9 km   30 km   90 km   300 km

6   **4.** Harriet Hare and Turbo Tortoise want to cross the finish line together on their 12 mile woodland race. Turbo sets off at 8:15 am and trots at a constant speed of 4 mph.

Harriet runs at a constant speed of 8 mph.

At what time should Harriet set off?

9   **5.** The world's fastest tortoise is acknowledged to be a leopard tortoise from County Durham called Bertie. In July 2014, Bertie sprinted along a 5.5 m long track in an astonishing 19.6 s.

Which of the following was Bertie's approximate average speed in kilometres per hour?

0.1   0.5   1   5   10

12   **6.** One gallon of honey provides enough fuel for a bee to fly about seven million miles.

Which of the following is roughly the number of bees that can fly one thousand miles when they have ten gallons of honey to share between them?

7000   70 000   700 000   7 000 000   70 000 000

12   7. Timmy, Tammy and Tommy all have tummy ache! They all set off
separately to visit their doctor, leaving their homes at exactly the same
time.

> Timmy cycles the 8 km to the surgery at an average speed of
> 20 km/h;
> Tammy walks the 1.2 km to the surgery at an average speed of
> 4 km/h; and
> Tommy drives the 16.5 km to the surgery at an average speed of
> 45 km/h.

In what order do they arrive at the surgery?

13   8. Alex Erlich and Paneth Farcas shared an opening rally of 2 hours and
12 minutes during their table tennis match at the 1936 World Games.
Each player hit around 45 shots per minute.

Which of the following is closest to the total number of shots played
in the rally?

200   2000   8000   12 000   20 000

14   9. In 2014 the *Tour de France* started in Leeds on 5 July. The year before,
the total length of the Tour was 3404 km and the winner, Chris Froome,
took a total time of 83 hours 56 minutes 40 seconds to cover this
distance.

Which of the following is closest to his average speed over the whole
event?

32 km/h   40 km/h   48 km/h   56 km/h   64 km/h

14   10. In a sponsored "Animal Streak" the cheetah ran at 90 km/h while the
snail slimed along at 20 h/km. The cheetah kept going for 18 seconds.

How long did the snail take to cover the same distance?

Areas

Exercise 14

5 **1.** The diagram shows a rectangle placed on a grid of 1 cm × 1 cm squares.

What is the area of the rectangle?

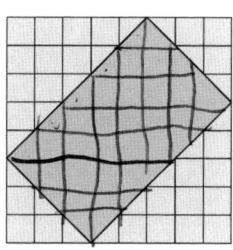

6 **2.** The absorptive surface of the small intestine of a typical adult has an area of approximately 200 square metres.

Which of the following has roughly the same area?

a dartboard	a table tennis table	a football pitch
a golf course	a tennis court	

7 **3.** Which of the following five straight lines cuts the shaded area in half?

XA   XB   XC   XD   XE

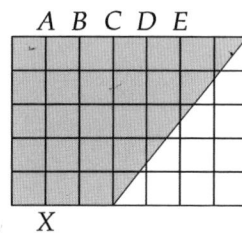

7   **4.** What area of this pennant is shaded grey?

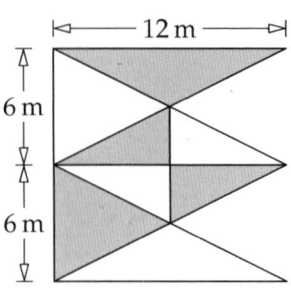

8   **5.** The diagram shows a figure made from six touching 1 × 1 squares arranged with a vertical line of symmetry. A straight line is drawn through the bottom corner *P* in such a way that the area of the figure is halved.

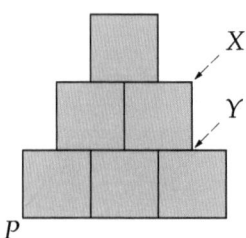

   At what distance from *X* does the line cross the side *XY*?

9   **6.** Which of the following five shaded regions has an area different from all the other shaded regions?

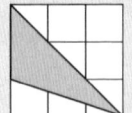 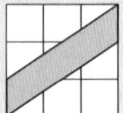 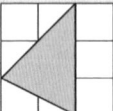 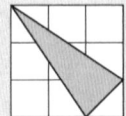 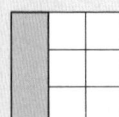

12   **7.** This figure is made from a straight line 16 cm long and two quarter-circles, one with its centre at the midpoint of the straight line.

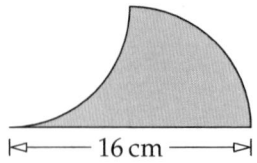

   What is the area of the figure?

13 **8.** What is the area of the shaded region in the rectangle?

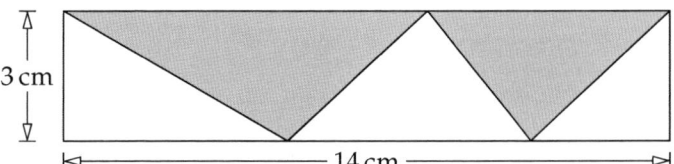

15 **9.** The circle has radius 1 cm. Two vertices of the square lie on the circle. One side of the square goes through the centre of the circle, as shown.

What is the area of the square?

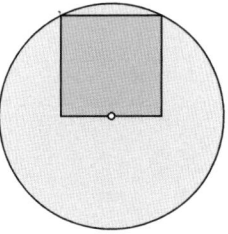

Three dimensions

Exercise 15

5 **1.** A solid 'star' shape is created by gluing a square-based pyramid, in which each edge is of length 1, onto each face of a cube of edge-length 1, so that the square base of each pyramid fits precisely onto a face of the cube.

How many faces does this 'star' have?

6 **2.** Four of the following five shapes can be placed together to make a cube.

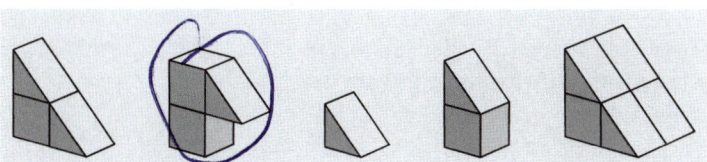

Which is the odd one out?

7 **3.** What is the product of the number of edges and the number of vertices of a tetrahedron?

[*A tetrahedron is a solid with four faces, each of which is a triangle.*]

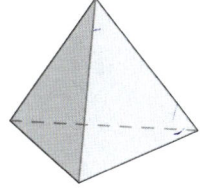

7   **4.** The faces of a regular octahedron are to be
      painted so that no two faces which have an
      edge in common are painted in the same
      colour.

      What is the smallest number of colours
      required?

      *[A regular octahedron is a solid with eight faces, each
      of which is an equilateral triangle.]*

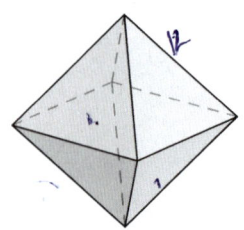

8   **5.** A large cube is made by stacking eight dice.
      The diagram shows the result, except that one
      of the dice is missing. Each die has faces with
      1, 2, 3, 4, 5 and 6 pips and the total number of
      pips on each pair of opposite faces is 7. When
      two dice are placed face to face, the matching
      faces must have the same number of pips.

      Which of the following could be the missing
      die?

  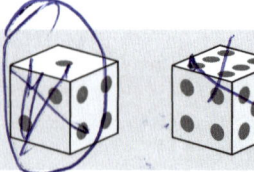

8   **6.** A square piece of card has a square of
      side-length 2 cm cut out from each of its
      corners. The remaining card is folded along
      the dashed lines shown to form an open box
      whose total internal surface area is 180 cm^2.

      What is the volume of the open box?

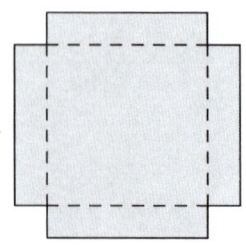

9   **7.** A solid wooden cube is painted blue on the outside. The cube is then
      cut into 27 smaller cubes of equal size.

      What fraction of the total surface area of these new cubes is blue?

9 **8.** The following five cuboids all have the same volume.
Which of them has the greatest surface area?

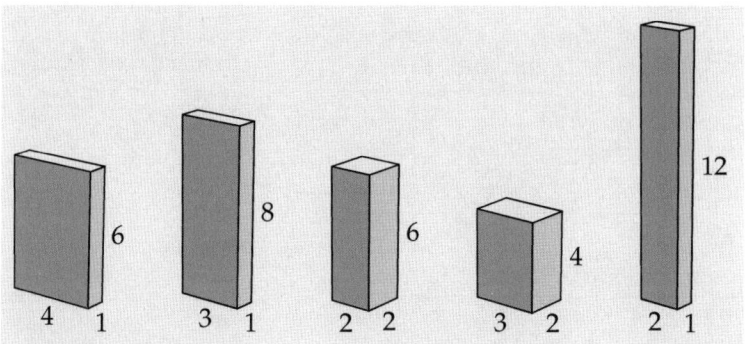

9 **9.** The diagram shows a net for a cube with a number
on each face. When the cube is made, three faces
meet at each vertex. The numbers on the three faces
which meet at each vertex are multiplied together.

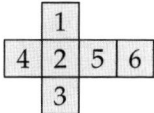

What is the largest product obtained?

11 **10.** The net shown on the left consists of squares and equilateral triangles.

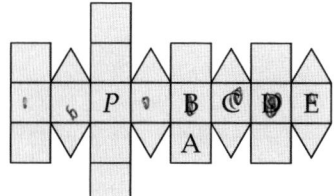
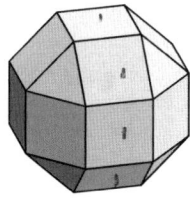

The net folds to form a rhombicuboctahedron, shown on the right.

When the face marked *P* is placed face down on a table, which face
will be facing up?

12 **11.** An ant is on the square marked with a black dot. The ant moves across an edge from one square to an adjacent square four times and then stops.

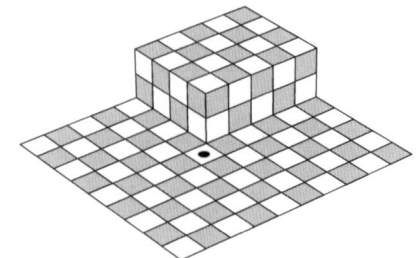

How many of the possible finishing squares are grey?

12 **12.** A cuboid is cut away from a cube of edge-length 10 cm as shown.

By what fraction does the total surface area of the solid decrease as a result?

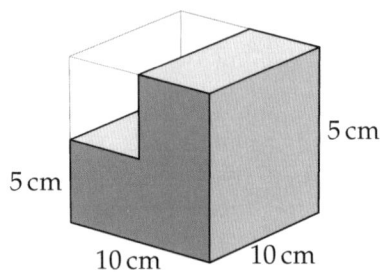

12 **13.** The sculpture '*Cubo Vazado*' [Emptied Cube] by the Brazilian artist Franz Weissmann is formed by removing cubical blocks from a solid cube to leave the symmetrical shape shown. All the edges have length 1, 2 or 3.

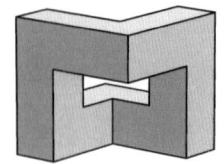

What is the surface area of the sculpture?

12 **14.** When a solid cube is held up to the light, how many of the following shapes could its shadow have?

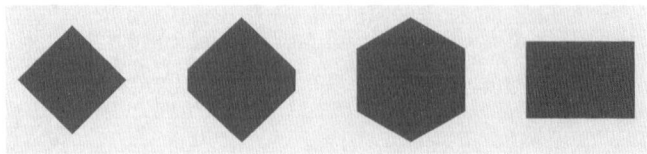

12 **15.** Each edge of a cube is to be coloured either red or black, so that every face of the cube has at least one black edge.

What is the smallest possible number of black edges?

Averages

Throughout the following exercise, *average* refers to what is sometimes called the mean.

Exercise 16

2 **1.** A tourist attraction, which opens every day, needs 30 000 visitors per day on average to break even. Last week there were 120 000 visitors.

What is the number of visitors needed this week to break even over the two-week period?

4 **2.** The mean, median and mode of the five numbers in the box below are the same.

7	7	5	7	?

What is the missing number?

6 **3.** A shop advertises 'Buy one, get one at half price'.

For this offer, the average cost per item is the same as:

Two for the price of one	Three for the price of one
Three for the price of two	Four for the price of three
Five for the price of four	

8   **4.** The average of three numbers x, y and z is x.

       In terms of x, what is the average of y and z?

10   **5.** The average weight of five giant dates was 50 g. Kate ate one and the average weight of the four remaining dates was 40 g.

       What was the weight of the date that Kate ate?

10   **6.** On four tests, each marked out of 100, my average was 85.

       What is the lowest mark I could have scored on any one test?

11   **7.** Three different positive integers have an average of 7.

       What is the largest positive integer that could be one of them?

13   **8.** The average of a list of 64 numbers is 64. The average of the first 36 numbers is 36.

       What is the average of the last 28 numbers?

14   **9.** In a sequence, each term after the first two terms is the average of all the terms which come before that term. The first term is 8 and the tenth term is 26.

       What is the second term?

14   **10.** What value of x makes the average of the first three numbers in this list equal to the average of the last four?

$$15 \quad 5 \quad x \quad 7 \quad 9 \quad 17$$

15   **11.** What is the average of $1.\dot{2}$ and $2.\dot{1}$?

More integers

Exercise 17

7 **1.** Just one positive integer has exactly eight factors including 6 and 15. What is the integer?

7 **2.** The numbers 1, 2, 3, 4, 5, 6, 7, 8, 9, 10 are all multiplied together. How many zeros are at the end of the answer?

7 **3.** Last year a newspaper reported that Turkish football team Sarigol Municipality transferred four of its players in return for a fee of 225 sacks of cement, needed to repair their stadium.

At the same rate of exchange, how many sacks of cement would be the transfer fee for a full team of eleven players and one reserve?

7 **4.** Suppose that m and n are positive integers and $m^2 + 2 = n^3$. Which of the following is a possible value of m?

$$2 \qquad 3 \qquad 4 \qquad 5 \qquad 6$$

7   **5.** Someone's 'birthday-product' is obtained by multiplying the day of the month in which they were born by the number of the month in which they were born, and then multiplying the answer by the year in which they were born.

Five English queens and their birthdays are given in the following box. Which of them has the same birthday-product as someone born on 5 February 1998? $18 \times 2 \times 1518$

Mary I, 18 February 1518   Elizabeth I, 7 September 1533
Anne, 6 February 1665   Victoria, 24 May 1819
Elizabeth II, 21 April 1926

9   **6.** A male punky fish has 9 stripes and a female punky fish has 8 stripes. I count 86 stripes on the fish in my tank.

What is the ratio of male fish to female fish?

9   **7.** A Langford number is one in which each digit of the number occurs twice; the digits 1 are separated by one other digit, the digits 2 are separated by two others, and so on.

Which of the following is a Langford number?

~~12142334~~   41312432   ~~14132342~~   ~~32432141~~   31213244

10   **8.** What is the remainder when 22 × 33 × 55 × 77 is divided by 8?

10   **9.** What is the smallest abundant number?

[*An* abundant *number is a positive integer N such that the sum of the factors of N (excluding N itself) is greater than N.*]

10   **10.** Fussy Fiona wants to buy a new house but she doesn't like house numbers that are divisible by 3 or by 5.

All the houses numbered between 100 and 150 inclusive are for sale.

How many houses can Fiona choose from?

11   **11.** For which of the following five numbers is the sum of all its factors *not* equal to a square?

3   22   40   66   70

11 **12.** The numbers 72, 8, 24, 10, 5, 45, 36, 15 are grouped in pairs so that the product of each pair is the same.

Which number is paired with 10?

12 **13.** What is the sum of the first 2011 digits when $20 \div 11$ is written as a decimal?

12 **14.** Gill is nine years old this year so she was fascinated to discover that a positive integer is divisible by nine precisely when the sum of its digits is also divisible by nine.

The six-digit number '1*d*3456' is a multiple of nine.

What is the value of the digit *d*?

14 **15.** In this addition sum, different letters represent different non-zero digits.

What is the value of $a + w + a + y$?

$$\begin{array}{r} f\ l\ y \\ +\ f\ l\ y \\ +\ f\ l\ y \\ \hline a\ w\ a\ y \end{array}$$

15 **16.** I have a bag of coins. In it, one third of the coins are gold, one fifth of them are silver, two sevenths are bronze and the rest are copper. My bag can hold a maximum of 200 coins.

How many coins are in my bag?

Reflections, rotations and dissections

Exercise 18

3   **1.** Professor Rosseforp has an unusual clock. The clock shows the correct time at noon, but the hands move anticlockwise rather than clockwise. The clock is very accurate, however, so the hands move at the correct speeds.

You look in a mirror at the Professor's clock at 1:30 pm. Which of the following could you see?

6   **2.** The equilateral triangle and regular hexagon shown have perimeters of the same length.

What is the ratio of the area of the triangle to the area of the hexagon?

6   **3.** The shape shown on the right was assembled from three identical copies of one of the following five smaller shapes, without gaps or overlaps.

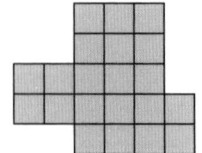

Which smaller shape was used?

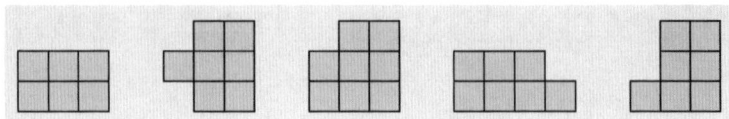

6   **4.** Triangle QRS is right-angled and isosceles. Beatrix reflects the letter P in the side QR to get an image. She reflects the first image in the side SQ to get a second image. Finally, she reflects the second image in the side RS to get a third image.

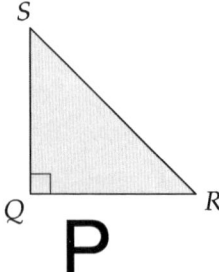

What does the third image look like?

8   **5.** The diagram shows a right-angled isosceles triangle XYZ, which circumscribes a square $PQRS$.

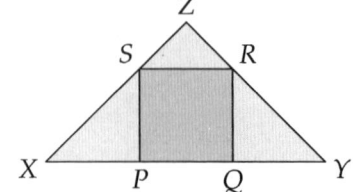

The area of triangle XYZ is x.

What is the area of square $PQRS$?

11   **6.** A square card printed with the letter N is held horizontally, as shown in the diagram, where the arrow indicates the direction of north.

The card is turned over by rotating it through $180°$ about an axis running from east to west, and then turned over by rotating it through $180°$ about an axis running from north-east to south-west.

After these rotations, how does the diagram on the card look to a person facing north?

12 **7.** The diagram shows a right-angled isosceles triangle divided into strips of equal width.

Four of the strips are shaded.

What fraction of the area of the triangle is shaded?

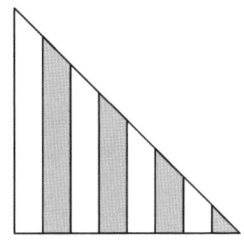

13 **8.** Four congruent trapeziums are placed so that their longer parallel sides form the diagonals of a square *PQRS*, as shown. The point *X* divides *PQ* in the ratio 3 : 1.

What fraction of the square is shaded?

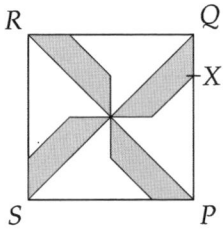

14 **9.** In the diagram, $\angle NOM = 130°$. The reflection of *OP* in *OM* is *OQ* and the reflection of *OP* in *ON* is *OR*.

What is the size of $\angle QOR$?

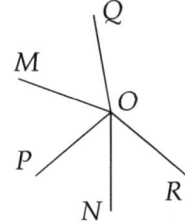

15 **10.** A flag is in the shape of a right-angled triangle, as shown, with the horizontal and vertical sides being of length 72 cm and 24 cm respectively. The flag is divided into 6 vertical stripes of equal width.

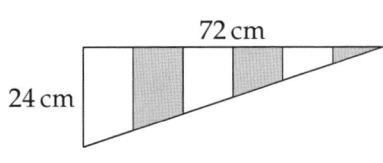

What is the difference between the areas of any two adjacent stripes?

15 **11.** The equilateral triangle ABC has sides of length 1 and AB lies on the line XY. The triangle is rotated clockwise around B until BC lies on the line XY. It is then rotated similarly around C and then about A as shown in the diagram.

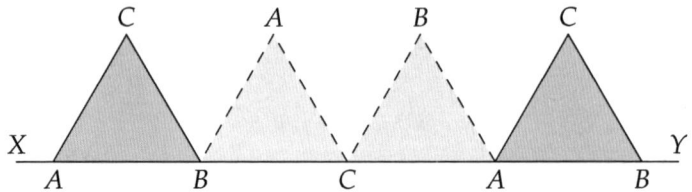

What is the length of the path traced out by point C during this sequence of rotations?

Algebra

Exercise 19

5   **1.** The square of a positive number is twice as big as the cube of that number.

What is the number?

5   **2.** The sum of two numbers is 2. The difference between them is 4.

What is their product?

7   **3.** The square of a non-zero number is equal to 70% of the original number.

What is the original number?

8   **4.** Jim rolled some dice and was surprised that the sum of the scores on the dice was equal to the product of the scores on the dice. One of the dice showed a score of 2, one showed 3 and one showed 5. The rest showed a score of 1.

How many dice did Jim roll?

9   **5.** At the age of twenty-six, Gill has passed her driving test and bought a car. Her car uses p litres of petrol per 100 km travelled.

How many litres of petrol would be required for a journey of d km?

10   **6.** Granny has taken up deep-sea fishing! Last week, she caught a fish
so big that she had to cut it into three pieces (head, body and tail) in
order to weigh it. The tail weighed 9 kg and the head weighed the
same as the tail plus one third of the body. The body weighed as much
as the head and tail together.

How much did the whole fish weigh?

10   **7.** Anna has 3 brothers and 5 sisters. Her brother Tom has S sisters and
B brothers.

What is the value of $S \times B$?

11   **8.** The standard Fibonacci sequence

$$1, 1, 2, 3, 5, 8, 13, \ldots$$

begins with two 1s, and each later number in the sequence is the sum
of the previous two numbers. Other Fibonacci-like sequences can be
constructed by starting with any two numbers a and b (not necessarily
1 and 1) and using the same rule for creating the other numbers in the
sequence.

What is the first term of the Fibonacci-like sequence whose second
term is 4 and whose fifth term is 22?

12   **9.** The sum of two numbers a and b is 7 and the difference between them
is 2.

What is the value of $a \times b$?

12   **10.** A long-sleeve shirt has 8 front buttons and 2 cuff buttons; a short-
sleeve shirt has 6 front buttons and no cuff buttons. The factory which
makes 'Slimboy Shirts' uses 10 times as many front buttons as cuff
buttons.

What is the ratio of long-sleeve shirts to short-sleeve shirts produced
by the factory?

12 **11.** Let $C°$ Celsius be the same temperature as $F°$ Fahrenheit, so that $F = \frac{9}{5} \times C + 32$.

To avoid working with fractions and awkward numbers, some people use the approximate formula $F \approx 2C + 30$.

What is the temperature in degrees Celsius when the approximate formula gives an answer which is too large by 1?

13 **12.** At Corbett's Ironmongery a fork handle and a candle cost a total of £6.10. The fork handle costs £4.60 more than the candle.

What is the cost of two fork handles and four candles?

14 **13.** Suppose that x, y and z satisfy the following three equations.

$$4x - y = 5$$
$$4y - z = 7$$
$$4z - x = 18$$

What is the value of $x + y + z$?

14 **14.** The integer p is positive and the integer q is negative.

Which of the following four expressions has the greatest value?

$$p - q \qquad q - p \qquad p + q \qquad -p - q$$

15 **15.** Zac halves a certain number and then adds 8 to the result. He finds that he obtains the same answer when he doubles his original number and then subtracts 8 from the result.

What is Zac's original number?

15 **16.** Wallace and Gromit are waiting in a queue. There are p people behind Wallace, who is q places in front of Gromit, and there are n people in front of Gromit.

How long is the queue?

Counting

Exercise 20

2 **1.** The information display on a train shows letters by illuminating dots in a rectangular 5×8 array.

In the letter t shown, what fraction of the dots in the array is illuminated?

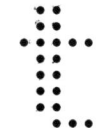

3 **2.** How many quadrilaterals are there in this diagram, which is constructed using 6 straight lines?

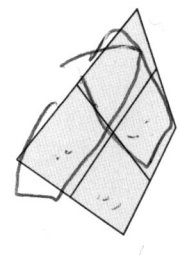

11 **3.** The diagram shows a square board in which strips of white cells alternate with strips of grey and white cells. A larger board, constructed in the same way, has 49 grey cells.

How many white cells are there on the larger board?

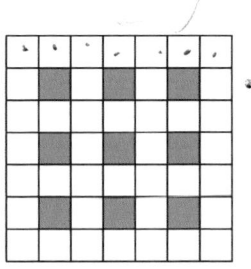

3 **4.** How many numbers can be written as a sum of two different positive integers each at most 100?

13   **5.** Pieces with the shape T are to be placed
      within the 5 × 5 grid shown, without
      overlapping, and with their edges along
      the lines of the grid.

      What is the maximum number of T-pieces
      that can be placed?

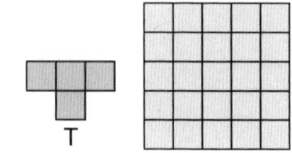

13   **6.** The diagram shows two rectangles which
      enclose five regions.

      What is the largest number of regions which
      can be enclosed by any two rectangles drawn
      on a sheet of paper?

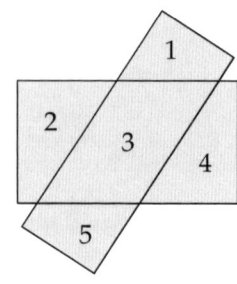

14   **7.** Sam is holding two lengths of rope by their midpoints. Pat chooses
      two of the loose ends at random and ties them together.

      What is the probability that Sam now holds one untied length of rope
      and one tied loop of rope?

14   **8.** A square patchwork quilt is made by joining four
      square pieces of cloth. Each piece is coloured grey
      and white, as shown. Only edges of the same colour
      are sewn together.

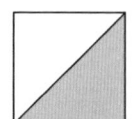

      How many different quilt patterns are possible? (Two patterns are
      considered to be the same if one can be rotated to look exactly like the
      other.)

15   **9.** In how many integers between 100 and 999 is the middle digit equal
      to the sum of the other two digits?

Circles and triangles

Exercise 21

7   **1.** Four touching circles all have radius 1 and their centres are at the corners of a square.

What is the radius of the circle through the points of contact X, Y, Z and T?

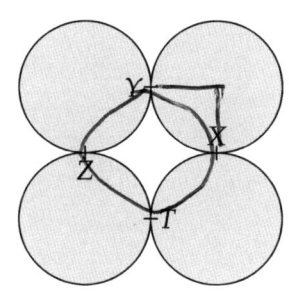

8   **2.** In the triangle PQR, PS = QS = RS. What is the size of angle QRP?

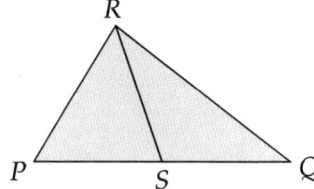

9   **3.** The perimeters of the three shapes shown are made up of straight
     lines and semicircular arcs of diameter 2. They fit snugly together, like
     jigsaw pieces.

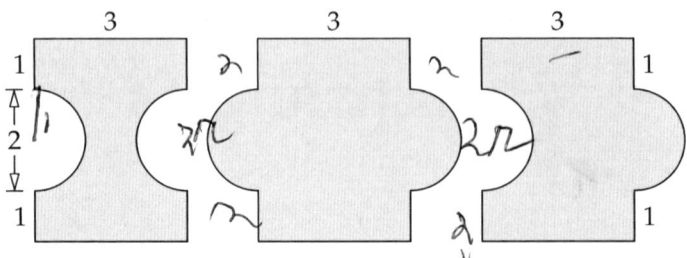

3   3   3

What is the difference between the total perimeter length of the three
separate pieces and the perimeter length of the shape formed when
the three pieces fit together?

10   **4.** The diagram shows five touching semicircles, each with radius 2.

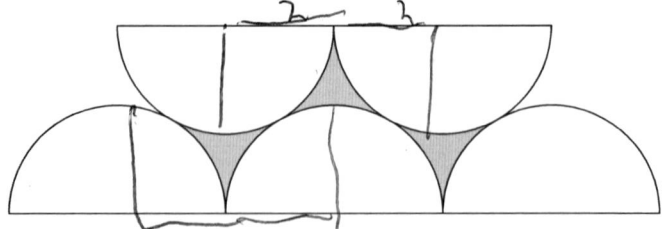

What is the length of the perimeter of the shaded shape?

10   **5.** Two sides of a triangle have lengths 6 cm and 5 cm. Perry suggests
     the following possible values for the length of the perimeter of the
     triangle:

     (i) 11 cm;
     (ii) 15 cm;
     (iii) 24 cm.

Which of Perry's suggestions could be correct?

11   **6.** Two of the sides of a right-angled triangle are 5 cm and 6 cm long.
     How many possibilities are there for the length of the third side?

12 **7.** In a right-angled triangle the two shorter sides have lengths 10 cm and 5 cm.

Which of the following is closest to the length of the hypotenuse?

11 cm   11.5 cm   12 cm   12.5 cm   13 cm

13 **8.** The diagram shows a rectangle with sides of length 5 cm and 4 cm.

Each arc is a quarter of a circle of radius 2 cm.

What is the total shaded area?

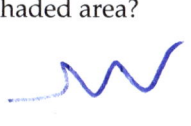

1 cm   2 cm   4 cm   2 cm

4 cm

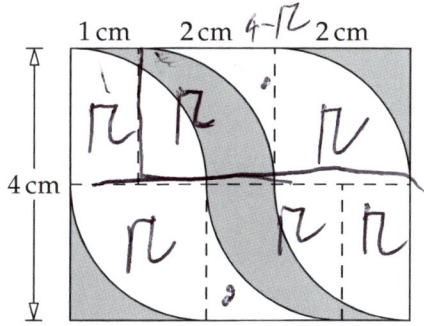

13 **9.** The diagram shows a semicircle containing a circle which touches the circumference of the semicircle and goes through its centre.

What fraction of the semicircle is shaded?

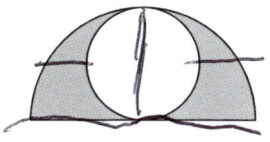

14 **10.** A 3 × 8 rectangle is cut into two pieces along the dotted line shown. The two pieces are rearranged to form a right-angled triangle.

What is the length of the perimeter of the triangle formed?

8

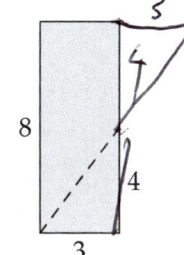

3

14 **11.** I have four rectangular pieces of thin hardboard whose dimensions
        are as follows.

$$55\,\text{cm} \times 85\,\text{cm}$$
$$65\,\text{cm} \times 75\,\text{cm}$$
$$65\,\text{cm} \times 85\,\text{cm}$$
$$90\,\text{cm} \times 105\,\text{cm}$$

Without bending the hardboard, how many of these can I get through
an open rectangular window measuring 60 cm × 80 cm?

15 **12.** Only one of the following five triangles can actually be made. Which
        one?

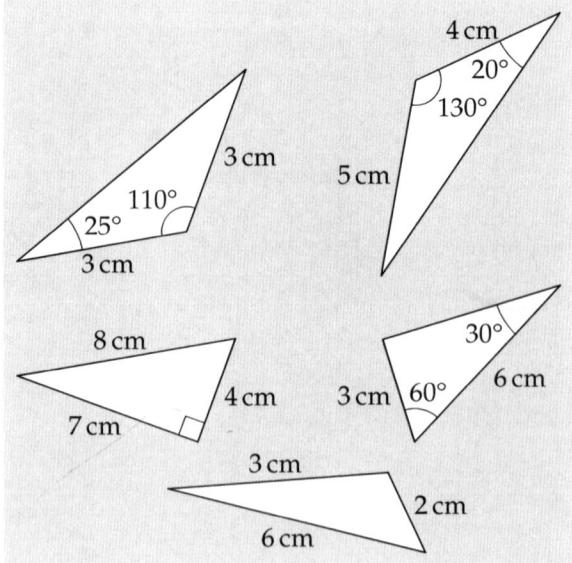

15   **13.** A pencil *PQ* lying on a table is given a half-turn about the end *Q* (so that *P* moves to *P'*) and then a half-turn about *P'* (so that *Q* moves to *Q'*). The point *R* on the pencil is one third of the way from *P* to *Q*.

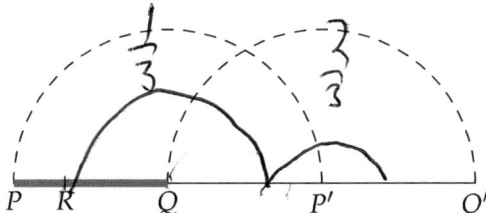

What is the ratio of the total distances moved by *P* and by *R*?

Part II

More challenging problems

More fractions, ratios and percentages

Exercise 22

10   **1.** What is the average of $\frac{1}{2}, \frac{1}{3}, \frac{1}{4}$ and $\frac{1}{6}$?

11   **2.** What is the value of $19\frac{1}{2} \times 20\frac{1}{2}$?

11   **3.** Which of the following five fractions is in the middle when they are written in numerical order?

$$\frac{4}{7} \qquad \frac{5}{8} \qquad \frac{3}{4} \qquad \frac{7}{11} \qquad \frac{8}{13}$$

13   **4.** The three blind mice stole a piece of cheese. In the night, the first mouse ate $\frac{1}{3}$ of the cheese. Later, the second mouse ate $\frac{1}{3}$ of the remaining cheese. Finally, the third mouse ate $\frac{1}{3}$ of what was then left of the cheese.

What fraction of the cheese did they eat between them?

14   **5.** Which of the following five expressions has the greatest value?

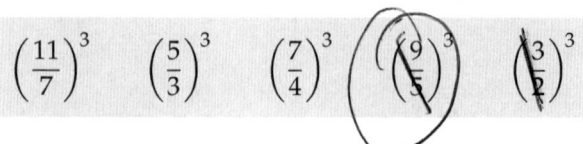

$$\left(\frac{11}{7}\right)^3 \quad \left(\frac{5}{3}\right)^3 \quad \left(\frac{7}{4}\right)^3 \quad \left(\frac{9}{5}\right)^3 \quad \left(\frac{3}{2}\right)^3$$

14   **6.** The ratio $a : b$ is equal to $2 : 3$ and the ratio $a : c$ is equal to $3 : 4$.
      What is the value of the ratio $b : c$?

15   **7.** Which of the following five calculations gives a value that is closest
      to 0?

$$\frac{1}{2}+\frac{1}{3}\times\frac{1}{4} \qquad \frac{1}{2}+\frac{1}{3}\div\frac{1}{4} \qquad \frac{1}{2}\times\frac{1}{3}\div\frac{1}{4} \qquad \frac{1}{2}-\frac{1}{3}\div\frac{1}{4}$$

$$\frac{1}{2}-\frac{1}{3}\times\frac{1}{4}$$

15   **8.** Suppose that $\dfrac{1}{x} = 3.5$.

      What is the value of $\dfrac{1}{x+2}$?

15   **9.** On the first day after the flood, half of Noah's animals escaped. On
      the second day one third of the remainder wandered off. On the third
      day one quarter of the rest hopped it.
      What fraction of Noah's original menagerie was left?

16   **10.** The base of a triangle is increased by 25% but the area of the triangle
       is unchanged.
       By what percentage is the corresponding height decreased?

16   **11.** The first two terms of a sequence are $\dfrac{2}{3}$ and $\dfrac{4}{5}$. Each term after the
       second term is the average of the two previous terms.
       What is the fifth term in the sequence?

16   **12.** After a year's training, Minnie Midriffe increased her average speed
       in the London Marathon by 25%.
       By what percentage did her time decrease?

16 **13.** I made just enough sticky treacle mixture to exactly fill a square tin with sides of length 12 inches. But all I could find were two $8\frac{1}{2}$ inch square tins.

How well would the mixture fit?

17 **14.** A shop advertised "Everything half price in our sale", but also now advertises that there is "An additional 15% off sale prices".

Overall, this is equivalent to what reduction on the original prices?

17 **15.** On the television programme *Antiques Roadshow*, a painting was said to be worth £15 000 although the painting had originally cost only 50p.

The painting was sold for £15 000. Which of the following gives the approximate percentage profit on the original cost?

| 15 000 | 30 000 | 300 000 | 1 500 000 | 3 000 000 |

20 **16.** The populations of five cities A, B, C, D, E in 1988 and 1998 are shown on the following scales.

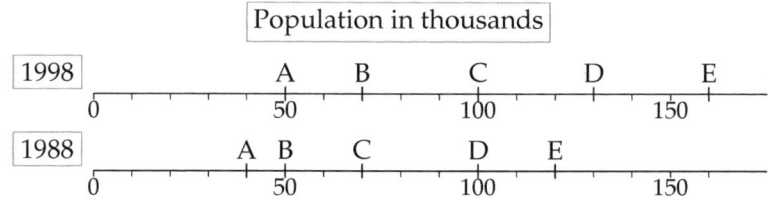

Population in thousands

Which of the five cities showed the largest percentage increase in population from 1988 to 1998?

21 **17.** For which integers n is the value of the product

$$\left(1+\frac{1}{2}\right)\left(1+\frac{1}{3}\right)\left(1+\frac{1}{4}\right)\cdots\left(1+\frac{1}{n}\right)$$

equal to an integer?

22   **18.** Inspector Remorse had a difficult year in 2004. A crime wave in Camford meant that he had 20% more cases to solve than in 2003, but his success rate dropped. In 2003, he solved 80% of his cases, but in 2004 he solved only 60% of them.

What was the percentage change in the number of cases he solved in 2004 compared with 2003?

25   **19.** Suppose that

$$x = \frac{111\,110}{111\,111}, \quad y = \frac{222\,221}{222\,223} \quad \text{ans} \quad z = \frac{333\,331}{333\,334}.$$

Which of the following five statements is correct?

$$x < y < z \qquad x < z < y \qquad y < z < x \qquad z < x < y \qquad y < x < z$$

Graphs and coordinates

Exercise 23

1. Which of the following could be the graph showing the circumference C of a circle in terms of its diameter D?

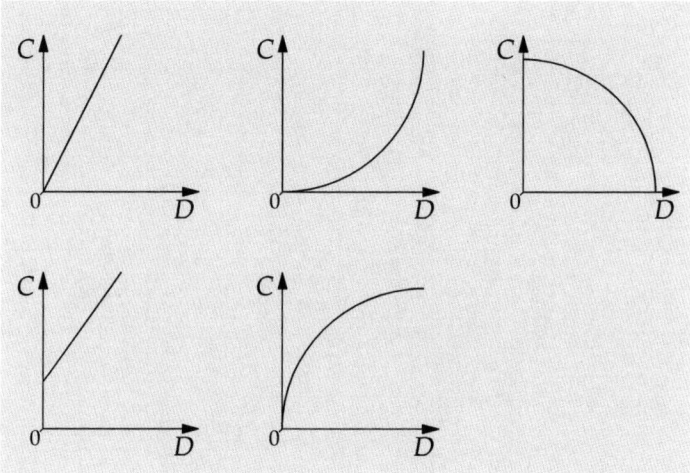

2. Four of the following points lie on a single straight line. Which is the odd one out?

$$(-3, -3) \quad (-2, -1) \quad (2, 5) \quad (4, 11) \quad (5, 13)$$

15   **3.** Bill is trying to sketch the graph of $y = 2x + 6$ but in drawing the axes
      he has placed the x-axis up the page and the y-axis across the page.

      Which of the following five graphs is a correct sketch of $y = 2x + 6$
      when the axes are placed in this way?

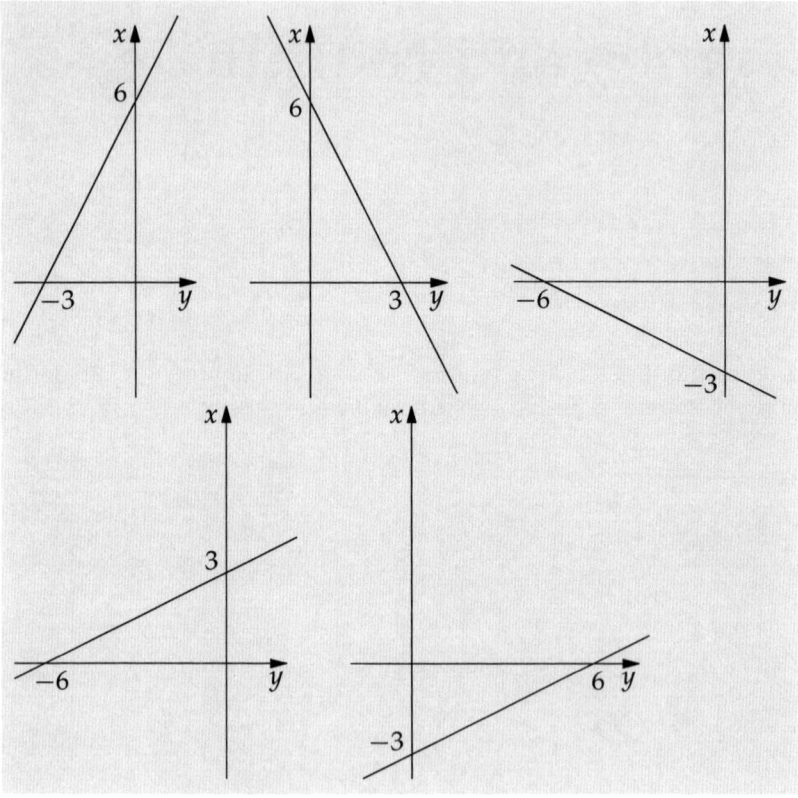

16   **4.** Which diagram shows the graph of $y = x$ after it has been rotated 90° clockwise about the point $(1, 1)$?

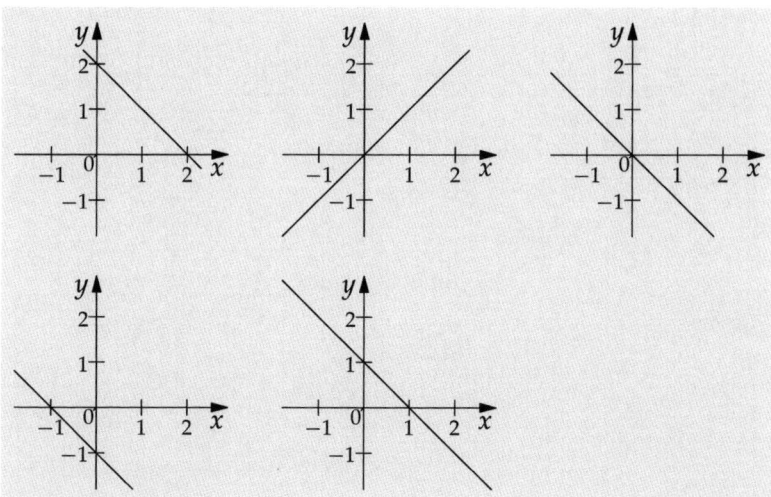

18   **5.** Consider looking from the origin $(0, 0)$ towards all the points (m, n), where each of m and n is an integer. Some points are *hidden*, because they are directly in line with another nearer point. For example, $(2, 2)$ is hidden by $(1, 1)$.

How many of the points $(6, 2)$, $(6, 3)$, $(6, 4)$, $(6, 5)$ are *not* hidden points?

18   **6.** Peri the winkle starts at the origin and slithers anticlockwise around a semicircular arc with centre $(4, 0)$. Peri then slides anticlockwise around a second semicircular arc with centre $(6, 0)$, and finally clockwise around a third semicircular arc with centre $(3, 0)$.

Where does Peri end this expedition?

23   **7.** A triangle is formed by the three straight lines with the following
        three equations.

$$y - x = 6$$
$$x - 2y = 3$$
$$x + y = 6$$

What is the area of the triangle?

More time and dates

Exercise 24

1. Albert Einstein is experimenting with two unusual clocks which both have 24-hour displays. One clock goes at twice the normal speed. The other clock goes backwards, but at the normal speed. Both clocks show the correct time at 13:00.

   What is the correct time when the displays on the clocks next agree?

2. How many weeks are there in $8 \times 7 \times 6 \times 5 \times 4 \times 3 \times 2 \times 1$ minutes?

3. Grannie's twelve-hour watch gains 30 minutes every hour, whilst Grandpa's twelve-hour watch loses 30 minutes every hour. At midnight, they both set their watches to the correct time of 12 o'clock.

   What is the correct time when their two watches next agree?

4. Albert Einstein was standing on the station platform thinking about relativity when he noticed that he could see two station clocks. Each clock was digital, showing only hours and minutes. He observed that the display on one clock changed to the next minute 10 seconds before the correct time, whereas the display on the other clock changed to the next minute 10 seconds after the correct time.

   For what fraction of the day did the clocks show the same time?

18   **5.** When dates are written using eight digits then 20 February 2002 is a palindromic date, because 20 02 2002 has the same digits in the same order when read in reverse. The previous palindromic date and the next few all occurred in the month of February.

What is the next month after February 2002, other than February, that had a palindromic date in it?

19   **6.** The product of Mary's age in years on her last birthday and her age now in complete months is 1800.

How old was Mary on her last birthday?

21   **7.** The year 2002 started on a Tuesday.

In which of the following years does each date fall on the same day of the week as it fell in 2002?

2008   2009   2012   2013   2014

23   **8.** Beatrix has a 24-hour digital clock on a glass table-top next to her desk. When she looked at the clock at 13:08, she noticed that the reflected display also read 13:08, as shown.

How many times in a 24-hour period do the display and its reflection give the same time?

24   **9.** A digital clock uses two digits to display hours, two digits to display minutes and two digits to display seconds, for example, 10:23:42.

How many times between 10:00:00 and 11:00:00 on the same morning are all six digits different?

More rates

Exercise 25

16   **1.** In 1967 Mike McNamara cycled 445.0 km in 12 hours, a new British men's record at that time. In the same time trial Beryl Burton (who died in 1996) completed 446.2 km.

How many more metres per hour was Beryl Burton's average speed than Mike McNamara's?

17   **2.** I walk to the bike shop at 3 miles per hour and cycle back along the same route at 12 miles per hour.

What is my average speed for the time I am actually travelling on the route?

18   **3.** Supergran walks from her chalet to the top of the mountain. She knows that if she walks at a speed of 6 mph she will arrive at 1 pm, whereas if she leaves at the same time and walks at 10 mph, she will arrive at 11 am.

At what speed should she walk if she wants to arrive at 12 noon?

18   **4.** An athlete covers three consecutive miles by walking the first mile, running the second and cycling the third. He runs twice as fast as he walks, and he cycles one and a half times as fast as he runs. He takes ten minutes longer than he would do if he cycled the three miles.

How long does he take by walking, running and cycling?

19 **5.** Driving to Birmingham airport, Mary cruised at 55 miles per hour for the first two hours and then flew along at 70 miles per hour for the remainder of the journey. Her average speed for the entire journey was 60 miles per hour.

How long did Mary's journey to Birmingham Airport take?

20 **6.** Max and his dog Molly set out for a walk. Max walked up the road and then back down again, completing a six mile round trip. Molly, being an old dog, walked at half Max's speed.

When Max reached the end of the road, he turned around and walked back to the starting point, at his original speed. Part way back he met Molly, who then turned around and followed Max home, still maintaining her original speed.

How far did Molly walk?

21 **7.** In a leisure park there are three running tracks, all with the same start *S* and finish *F*, and all made from either one or two semicircular arcs with centres on the straight line *SF*.

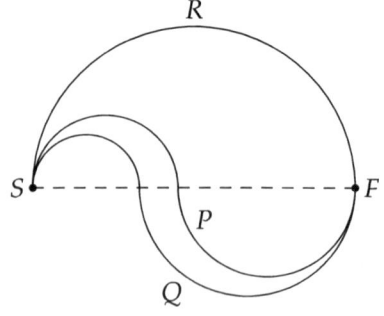

Three runners *P*, *Q* and *R* start together at *S* and run at the same constant speed along the tracks, as shown.

In what order do they reach *F*?

24 **8.** Every day, Aimee goes up an escalator on her journey to work. If she stands still, it takes her 60 seconds to travel from the bottom to the top.

One day the escalator was broken so she had to walk up it. This took her 90 seconds.

How many seconds would it take her to travel up the escalator if she walked up it, at the same speed as before, while it was working?

Factors and types of integers

Exercise 26

16   **1.** How many of the following four integers are divisible by 24?

$$2^2 \times 3^2 \times 5^2 \times 7^3 \qquad 2^2 \times 3^3 \times 5^2 \times 7^2$$
$$2^2 \times 3^3 \times 5^2 \times 7^2 \qquad 2^3 \times 3^2 \times 5^2 \times 7^2$$

16   **2.** You are asked to choose two positive integers m and n, with $m > n$, so that as many as possible of the expressions

$$m + n \qquad m - n \qquad m \times n \qquad m \div n$$

have values that are prime.

When you do this correctly, how many of these four expressions have values that are prime?

16   **3.** How many different positive integers n are there for which n and $n^3 + 3$ are both prime?

16 **4.** There are three statements in the box.

> (i) 3^{10} is even
> (ii) 3^{10} is odd
> (iii) 3^{10} is square

Exactly which ones are true?

18 **5.** The positive integers p and q are distinct primes less than seven.

What is the largest possible value of the highest common factor of $2p^2q$ and $3pq^2$?

18 **6.** The integers from 1 to 20 are listed below in such a way that the sum of each adjacent pair is prime. Missing numbers are indicated by the $*$ symbol.

> 20, $*$, 16, 15, 4, $*$, 12, $*$, 10, 7, 6, $\underline{*}$, 2, 17, 14, 9, 8, 5, 18, $*$.

Which number goes in the underlined place?

19 **7.** The following sequence continues indefinitely:

> $27 = 3 \times 3 \times 3$, $\quad 207 = 3 \times 3 \times 23$, $\quad 2007 = 3 \times 3 \times 223$,
> $20\,007 = 3 \times 3 \times 2223$, $\quad \ldots$.

One of the following five integers is a multiple of 81. Which one?

> 200 007 $\quad$ 20 000 007 $\quad$ 2 000 000 007 $\quad$ 200 000 000 007
> 20 000 000 000 007

19 **8.** Trinni is fascinated by the triangular numbers (1, 3, 6, 10, 15, 21, and so on).

Recently, coming across a clock, she found that she could rearrange the twelve numbers 1, 2, 3, ..., 12 around the face so that each adjacent pair added up to a triangular number. She left the 12 in its usual place.

What number did she put where the 6 would usually be?

20 **9.** One of the following five numbers is the largest of nine consecutive positive integers whose sum is a square.

Which one is it?

118   128   138   148   158

20 **10.** What is the largest power of two that divides $127^2 - 1$?

22 **11.** The positive integers a, b and c are all different. None of them is a square but all the products ab, ac and bc are squares.

What is the least value that $a + b + c$ can take?

23 **12.** As n takes each positive integer value in turn (that is, $n = 1$, $n = 2$, $n = 3$, and so on) how many different values are obtained for the remainder when n^2 is divided by $n + 4$?

23 **13.** For how many values of n are both n and $\dfrac{n+3}{n-1}$ integers?

24 **14.** Given any positive integer n, Paul adds together the distinct factors of n, other than n itself.

Which of the following five numbers can never be Paul's answer?

1   3   5   7   9

More in three dimensions

Exercise 27

1. A wooden cube with edge-length 12 cm is cut up into small cubes with edge-length 1 cm.

Which of the following five expressions is equal to the total length of all the edges of all these small cubes?

$$12 \,\text{cm} \qquad 12^2 \,\text{cm} \qquad 12^3 \,\text{cm} \qquad 12^4 \,\text{cm} \qquad 12^5 \,\text{cm}$$

2. A $1 \times 2 \times 3$ block is placed on an 8×8 board, as shown, with the 1×2 face X at the bottom. It is rolled over an edge without slipping onto the 1×3 face Y, then onto the 2×3 face Z, then onto X, Y, Z again in that order.

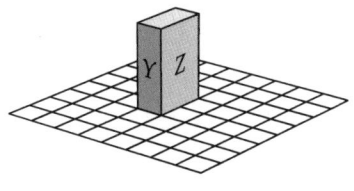

How many different cells on the board has the block occupied altogether, including the starting and ending positions?

17   **3.** The football shown is made by sewing together
      12 black pentagonal panels and 20 white
      hexagonal panels. There is a join wherever two
      panels meet along an edge.

      How many joins are there?

17   **4.** Last year Gill's cylindrical twenty-first birthday cake wasn't big
      enough to feed all her friends. This year she will double the radius
      and triple the height.

      What will be the ratio of the volume of this year's birthday cake to the
      volume of last year's cake?

19   **5.** A snail is at one corner of the top face of a cube with edge-length 1 m.
      The snail can crawl at a speed of 1 m per hour.

      What proportion of the cube's surface is made up of points which the
      snail could reach within one hour?

19   **6.** Three rectangular-shaped holes have been drilled passing all the way
      through a solid 3 × 4 × 5 cuboid. The diagrams show the front, side
      and top views of the resulting block.

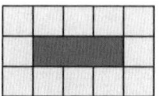 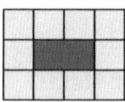 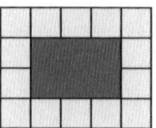

      What fraction of the original cuboid remains?

19 **7.** A can has the shape of a circular cylinder. The can contains lemonade, shown shaded in the diagram, in which *XY* is a diameter.

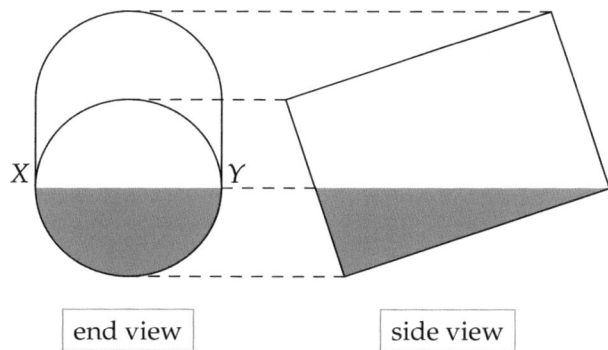

| end view | side view |

Which of the following describes the fraction of the volume of the can that is filled with lemonade?

just below a quarter   just above a quarter   exactly a quarter
just below a half   exactly a half

20 **8.** The total length of the edges of a cube is L cm. The surface area of the cube is L cm^2.

What is the volume of the cube?

22 **9.** The diagram shows an ordinary die. It is placed on a horizontal table with the '1' face facing east.

The die is moved four times, rotating it each time through 90° about an edge. The faces in contact with the table are first 1, then 2, then 3, then 5.

In which direction is the '1' face facing after this sequence of moves?

[In an ordinary die, the scores on opposite faces always total 7.]

22 **10.** A regular tetrahedron with edges of length
6 cm has each corner cut off to produce the
solid shown. The removed corners are all
regular tetrahedra, but not necessarily all the
same size.

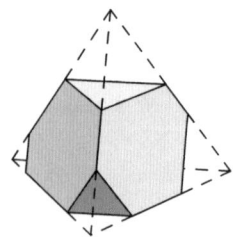

What is the total length of the edges of the
resulting solid?

*[A regular tetrahedron is a solid with four faces, each of which is an equilateral
triangle.]*

24 **11.** The diagram shows a cube of edge-length 1 on
which all twelve face diagonals have been drawn,
creating a network with 14 vertices (the original
eight corners, plus the six face centres) and 36
edges (the original twelve edges of the cube plus
four extra edges on each face).

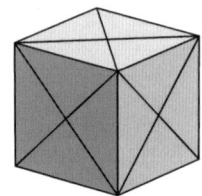

What is the length of the shortest path along the edges of the network
which passes through all 14 vertices?

24 **12.** How many space diagonals are there in a regular
dodecahedron?

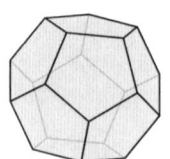

*[A dodecahedron is a polyhedron with twelve faces; each face
of a regular dodecahedron is a regular pentagon, as shown.*

*A space diagonal of a polyhedron is a line segment joining
two vertices of the polyhedron that do not lie in the same face.]*

More algebra

Exercise 28

1. Pat, Quentin, Robin and Sam have sums of money totalling £150. Pat and Quentin have £55 between them and Pat and Robin have £65 between them.

   What is the difference between the amounts that Pat and Sam have?

2. The total weight of a box, 20 plates and 30 cups is 4.8 kg. The total weight of the box, 40 plates and 50 cups is 8.4 kg.

   What is the total weight of the box, 10 plates and 20 cups?

3. Seventy pupils—37 boys and 33 girls—are divided into two groups, with forty pupils in Group I and thirty pupils in Group II.

   How many more boys are there in Group I than there are girls in Group II?

4. Seven rolls weigh the same as four crumpets. Five scones weigh the same as six crumpets. Each crumpet weighs c grams, each roll weighs r grams, and each scone weighs s grams.

   Which of the following five statements is true?

   $$r < s < c \qquad s < r < c \qquad s < c < r \qquad c < s < r \qquad r < c < s$$

19   **5.** Which of the following five expressions is equivalent to

$$\left(x \div (y \div z)\right) \div \left((x \div y) \div z\right)?$$

$$1 \qquad \frac{1}{xyz} \qquad x^2 \qquad y^2 \qquad z^2$$

19   **6.** Which of the following five expressions is equal to

$$(1 + x + y)^2 - (1 - x - y)^2$$

for all values of x and y?

$$4x \qquad 2(x^2 + y^2) \qquad 0 \qquad 4xy \qquad 4(x + y)$$

19   **7.** Four wiggles are the same as three woggles; two woggles are the same as five waggles; and six waggles are the same as one wuggle.

Which of the following four expressions has the smallest value?

1 wuggle   2 woggles   3 waggles   4 wiggles

20   **8.** A rectangle with area $125 \, \text{cm}^2$ has sides in the ratio $4 : 5$.

What is the length of the perimeter of the rectangle?

20   **9.** Inspector Remorse estimates that he can solve the average murder in h hours, a bank robbery in half that time, and a car theft in one third of the time he takes to solve a bank robbery.

In terms of h, how many hours would Inspector Remorse expect to take in solving two murders, six car thefts and four bank robberies?

21 **10.** A square is divided into four congruent rectangles and a smaller square, as shown. The area of the small square is a quarter of the area of the whole square.

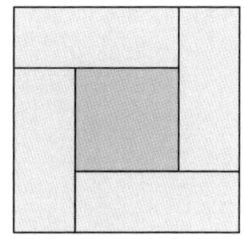

What is the ratio of the length of a short side of one of the rectangles to the length of a long side?

21 **11.** Suppose that a, b, c and d satisfy the following three equations.

$$a = b - c$$
$$b = c - d$$
$$c = d - a$$

What is the value of $\dfrac{a}{b} + \dfrac{b}{c} + \dfrac{c}{d} + \dfrac{d}{a}$?

22 **12.** In a mathematics examination with N questions, you score m marks for a correct answer to each of the first q questions and $m + 2$ marks for a correct answer to each of the remaining questions.

What is the maximum possible score?

22 **13.** In the *Soft Boulder Café* each table has three legs, each chair has four legs and all the customers and the three members of staff have two legs each. There are four chairs at each table. At a certain time, three-quarters of the chairs are occupied by customers and there are 206 legs altogether in the café.

How many chairs does the café have?

23 **14.** Suppose that it takes p men q hours to paint r square metres of the Forth Bridge.

How long would it take s men to paint t square metres of the bridge?

25 **15.** The width : height ratio of television screens is changing from the traditional 4 : 3 to the widescreen 16 : 9.

What is the ratio widescreen width : traditional width for a traditional screen and a widescreen with the same area?

(Assume that television screens are rectangles.)

More digits and integers

Exercise 29

.6 **1.** The Pythagoras Patisserie sells triangular cakes at 39p each and square buns at 23p each. For her party, Helen spent exactly £5.12 on an assortment of these cakes and buns.

How many items in total did Helen buy?

.6 **2.** The pattern

$$12345\ 12345\ 12345\ldots$$

is continued to form a two-thousand-digit number.
What is the sum of all 2000 digits?

.7 **3.** Baldrick can afford to buy either 6 parsnips and 7 turnips or 8 parsnips and 4 turnips. Both options leave him with no change whatsoever.

If, however, he bought only his beloved turnips, how many could he afford?

.8 **4.** The number $3^4 \times 4^5 \times 5^6$ is written out in full.

How many zeros are there at the end of the number?

.9 **5.** A list of positive integers has a median of 8, a mode of 9 and a mean of 10.

What is the smallest possible number of integers in the list?

19   **6.** The diagram shows a large rectangle composed of nine identical smaller rectangles. Both the length and breadth of each of these smaller rectangles are whole numbers of centimetres.

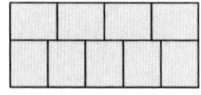

Which of the following five expressions could be the area of the large rectangle?

$$450\,\text{cm}^2 \qquad 630\,\text{cm}^2 \qquad 1260\,\text{cm}^2 \qquad 1440\,\text{cm}^2 \qquad 1620\,\text{cm}^2$$

20   **7.** Suppose that 2006 is the correct answer to the calculation

$$1 - 2 + 3 - 4 + 5 - 6 + \cdots + (n-2) - (n-1) + n.$$

What is the sum of the digits of n?

20   **8.** Suppose that e, i, n and t are different positive integers that satisfy the following three equations.

$$n + i + n + e = 9$$
$$t + e + n = 10$$
$$i = 1$$

What is the value of t?

21   **9.** In King Arthur's jousting tournament, each of the several competing knights receives 17 points for every bout he enters. The winner of each bout receives an extra 3 points. At the end of the tournament, the Black Knight has exactly one more point than the Red Knight.

What is the smallest number of bouts that the Black Knight could have entered?

22   **10.** What is the maximum possible value of the median number of cups of coffee bought per customer on a day when Sundollars Coffee Shop sells 477 cups of coffee to 190 customers, and every customer buys at least one cup of coffee?

22 **11.** At a cinema, a child's ticket costs £4.20 and an adult's ticket costs £7.70. When a group of adults and children went to see a film, the total cost was £*T*.

Which of the following five numbers is a possible value of *T*?

$$91 \quad 92 \quad 93 \quad 94 \quad 95$$

24 **12.** Jasmine spends exactly £120 on three types of plants: poisoned ivy, deadly nightshade and triffids. Poisoned ivy plants cost £2 each, deadly nightshade plants cost £9 each and triffids cost £12 each. She buys twenty plants in total, including at least one of each type.

How many triffids did she buy?

25 **13.** In 1984 the engineer and prolific prime-finder Harvey Dubner found the largest known prime each of whose digits is either a one or a zero. The prime can be expressed as

$$\frac{10^{641} \times (10^{640} - 1)}{9} + 1.$$

How many digits does this prime have?

25 **14.** Suppose that $5^j + 6^k + 7^\ell + 11^m = 2006$, where j, k, ℓ and m are different non-negative integers.

What is the value of $j + k + \ell + m$?

More circles

Exercise 30

1. The shaded region in the diagram, bounded by two concentric circles, is called an *annulus*. The circles have radii 2 cm and 14 cm.

The dashed circle divides the area of this annulus into two equal parts. What is its radius?

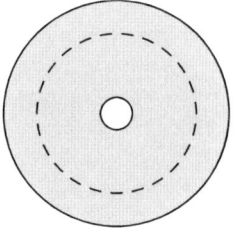

2. The diagram shows four equal discs and a square. Each disc touches its two neighbouring discs, and each corner of the square is positioned at the centre of a disc. The side-length of the square is $\frac{2}{\pi}$.

What is the length of the perimeter of the shaded region?

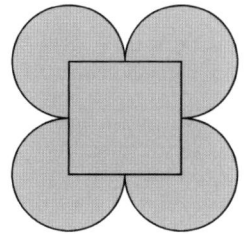

17   **3.** The shaded region is bounded by eight
     equal circles with centres at the corners and
     midpoints of the sides of a square.   The
     perimeter of the square has length 8.

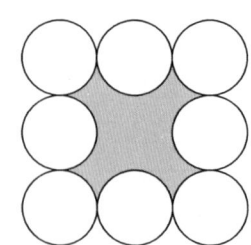

   What is length of the perimeter of the shaded
   region?

17   **4.** The diagram shows three semicircles,
     each of radius one.

   What is the area of the shaded region?

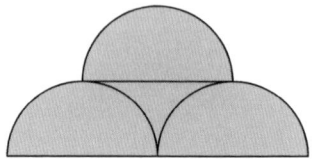

17   **5.** The three circles in the diagram have the
     same centre and have radii 3 cm, 4 cm and
     5 cm.

   What percentage of the area of the largest
   circle is shaded?

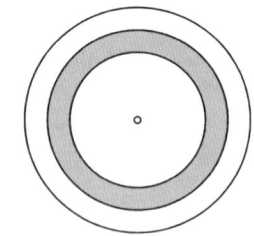

19   **6.** The shaded region shown in the diagram
     is bounded by four arcs, each of the same
     radius as that of the surrounding circle.

   What fraction of the surrounding circle is
   shaded?

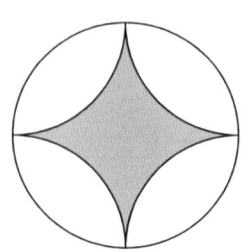

20   **7.** Two semicircles are drawn in a
     rectangle as shown.

   What is the width of the overlap of
   the two semicircles?

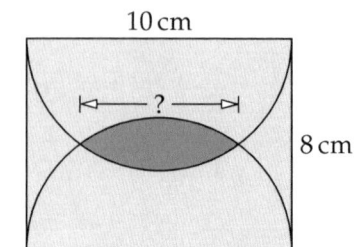

20 **8.** The diagram shows a regular pentagon and five circular arcs. The sides of the pentagon have length 4. The centre of each arc is a vertex of the pentagon, and the ends of the arc are the midpoints of the two adjacent sides.

What is the total area of the shaded regions?

20 **9.** P, Q, R are points on the circumference of a circle of radius 4 cm. $\angle PQR = 45°$.

What is the length of chord PR?

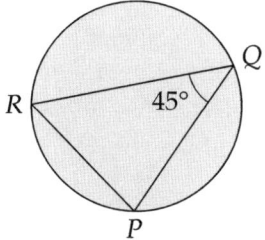

21 **10.** The diagram shows two circles and four equal semicircular arcs. The area of the inner shaded circle is 1.

What is the area of the outer circle?

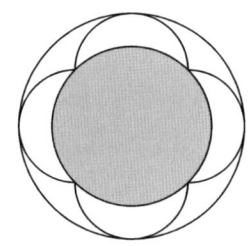

21 **11.** The diagram shows two semicircular arcs, PQRS and ROQ. The diameters PS and QR of the two semicircles are parallel; PS is of length 4 and is a tangent to semicircular arc ROQ.

What is the area of the shaded region?

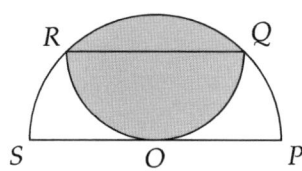

23   **12.** A Saxon silver penny, from the reign of Ethelred II in the eighth century, was sold in 2014 for £78 000.

A design on the coin depicts a circle surrounded by four equal quarter-circles, as shown. The width of the design is 2 cm.

What is the radius of the small circle?

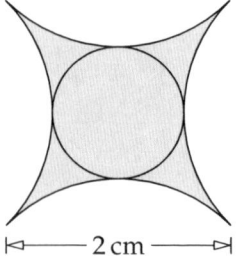

23   **13.** A sector of a disc is removed by making two straight cuts from the circumference to the centre. The perimeter of the sector has the same length as the circumference of the original disc.

What fraction of the area of the disc is removed?

23   **14.** A window frame in Salt's Mill consists of two equal semicircles and a circle inside a large semicircle, with each touching the other three, as shown. The width of the frame is 4 m.

What is the radius of the circle?

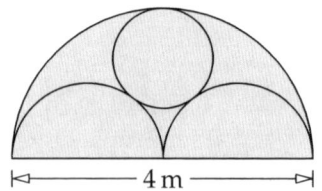

23   **15.** The square $PQRS$ is inscribed in a circle of radius 1. Semicircles are drawn with diameters PQ, QR, RS and SP as shown.

What is the total area of the four shaded regions?

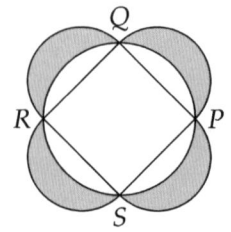

24   **16.** The diagram shows a shaded shape bounded by circular arcs with the same radius. The centres of three arcs are the vertices of an equilateral triangle; the other three centres are the midpoints of the sides of the triangle. The sides of the triangle have length 2.

What is the difference between the area of the shaded shape and the area of the triangle?

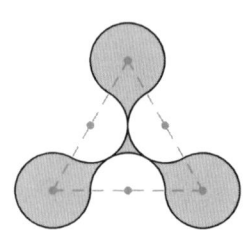

17. PQ is a diameter of a circle of radius $1\,\text{cm}$. Two circular arcs of equal radius are drawn with centres P and Q. These arcs meet on the circle, as shown.

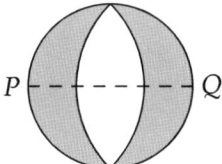

What is the shaded area?

Miscellany 2

Exercise 31

16 **1.** The pattern shown is continued.
What number will appear directly
below 400?

```
              1
           2   3   4
        5   6   7   8   9
     10  11  12  13  14  15  16
```

17 **2.** The first term of a sequence of positive integers is 6. The other terms in the sequence follow these two rules:

   (i) when a term is even then divide it by 2 to obtain the next term;

  (ii) when a term is odd then multiply it by 5 and subtract 1 to obtain the next term.

For which values of n is the nth term equal to n?

17 **3.** Platinum is a very rare metal, even rarer than gold. Its density is $21.45\,\mathrm{g/cm^3}$. The world production has been about 110 tonnes for each of the past 50 years, and negligible before that.

Which of the following has a comparable volume to that of the total amount of platinum ever produced?

> a shoe box   a cupboard   a house   Buckingham Palace
> the Grand Canyon

17   **4.** A, B, C, D, E, P and Q are points on the number line, as shown.

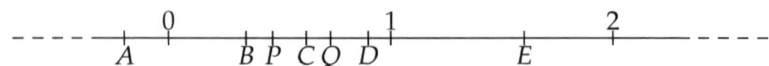

One of the points represents the product of the numbers represented by P and Q. Which is it?

18   **5.** In the calculation $1003 \div 4995 = 0.2\dot{0}0\dot{8}$, the number $0.2\dot{0}0\dot{8}$ represents the recurring decimal $0.2008008008008\ldots$

When the answers to the following five calculations are arranged in numerical order, which one is in the middle?

$$226 \div 1125 = 0.200\dot{8} \qquad 251 \div 1250 = 0.2008$$
$$497 \div 2475 = 0.20\dot{0}\dot{8} \qquad 1003 \div 4995 = 0.2\dot{0}0\dot{8}$$
$$2008 \div 9999 = 0.\dot{2}00\dot{8}$$

18   **6.** Suppose that $4^x + 4^x + 4^x + 4^x = 4^{16}$.

What is the value of x?

19   **7.** Suppose that $8^m = 27$.

What is the value of 4^m?

20   **8.** A square with sides of length 2 is folded in half to form a triangle. A second fold is made, parallel to the first, so that the apex of this triangle folds onto a point on its base, thereby forming a trapezium.

What is the length of the perimeter of this trapezium?

21   **9.** Suppose that x is positive and less than 1.

Which of the following five numbers is the largest?

$$x^2 + x \qquad x^2 \qquad x^3 \qquad x^3 + x^2 \qquad x^4$$

21 **10.** A wire in the shape of an equilateral
triangle with sides of length 9 cm is
placed flat on a piece of paper.

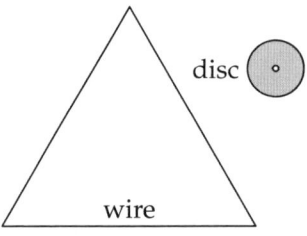

A pencil is held in the hole at the
centre of a disc of radius 1 cm, and
the disc is rolled all the way around
the outside of the wire, and then all
the way around the inside of the wire.

Which of the following five shapes does the pencil draw?

22 **11.** Suppose that $5^p = 9$, $9^q = 12$, $12^r = 16$, $16^s = 20$ and $20^t = 25$.

What is the value of $pqrst$?

22 **12.** Curly and Larry like to have their orange squash made to the same
strength. Unfortunately, Moe has put 25 ml of squash with 175 ml of
water in Curly's glass and 15 ml of squash with 185 ml of water in
Larry's glass.

Is it possible to put some of the mixture in Curly's glass into Larry's
glass so that they end up with drinks of the same strength? If so, how
many millilitres should be transferred?

22 **13.** The diagram shows an irregular
hexagon with interior angles all
equal to 120° made by cutting the
corners off a piece of card in the
shape of an equilateral triangle
with sides of length 20.

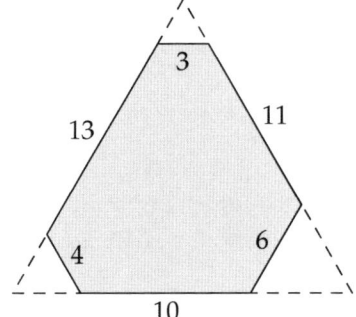

An identical hexagon could also be
made by cutting the corners off a
different equilateral triangle; what
is the side-length of this triangle?

22 **14.** For four of the following five shapes, four exact copies can be fitted together to make a rectangle.

Which is the odd one out?

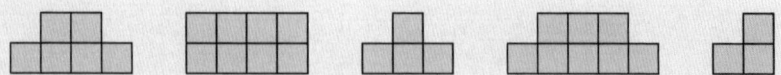

23 **15.** In the diagram, the letter S is made from two arcs *KL* and *MN*, and the line segment *LM*. Each arc is five-eighths of the circumference of a circle of radius 1, and *LM* is tangent to both circles. At points *K* and *N*, common tangents to the two circles touch one of the circles.

What is the length *LM*?

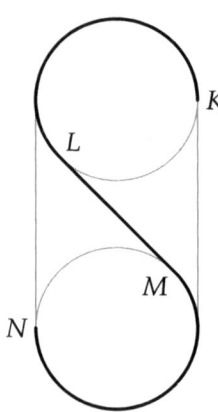

24 **16.** All the positive integers are written in the cells of a square grid. Starting from 1, the numbers spiral anticlockwise. The first part of the spiral is shown in the diagram.

What number will be immediately below 2012?

				...	32	31
	17	16	15	14	13	30
	18	5	4	3	12	29
	19	6	1	2	11	28
	20	7	8	9	10	27
	21	22	23	24	25	26

24 **17.** The integers p, q and $p - q$ are all positive, which of the following five expressions has least value?

$$\frac{q^2}{p^2} \qquad \frac{p^2}{q^2} \qquad \frac{q}{p} \qquad \sqrt{\frac{q}{p}} \qquad \sqrt{\frac{p}{q}}$$

18. The diagram shows a semicircle and an isosceles triangle which have equal areas.

What is the value of tan $x°$?

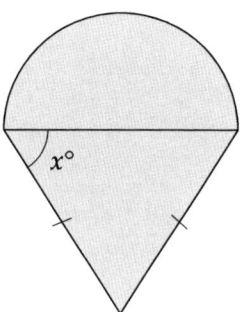

More angles and polygons

Exercise 32

1. The diagram shows a large equilateral triangle divided by three straight lines into seven regions. The three dark regions are equilateral triangles with sides of length 5 cm and the central white region is an equilateral triangle with sides of length 2 cm.

What is the side-length of the original large triangle?

2. The diagram shows a parallelogram inside a triangle. The marked lengths are equal.

What is the value of x?

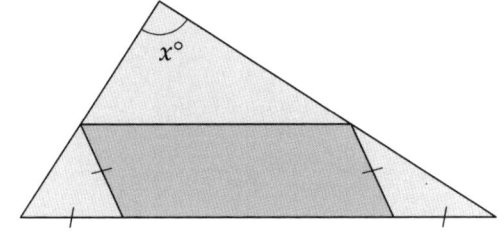

19   **3.** The diagram shows a regular pentagon
     and a regular hexagon which overlap.

     What is the value of x?

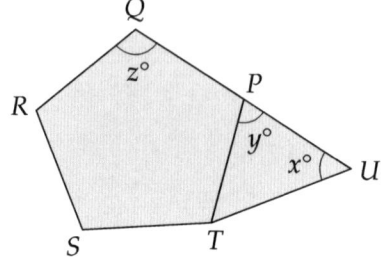

19   **4.** $PQRST$ is a regular pentagon.
     UPQ is a straight line and
     $UP = PQ$.

     What is the value of the ratio
     $x : y : z$?

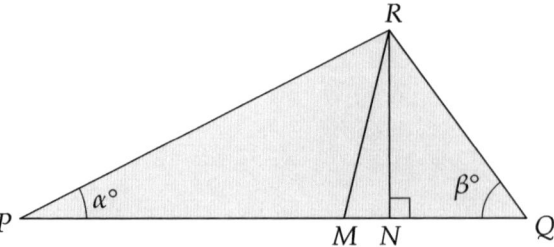

21   **5.** In the triangle PQR in the diagram below, $\angle RPQ = \alpha°$ and $\angle PQR = \beta°$, where $\alpha < \beta$. The points M and N lie on PQ so that RM bisects $\angle QRP$ and RN is the perpendicular from R to the line PQ.

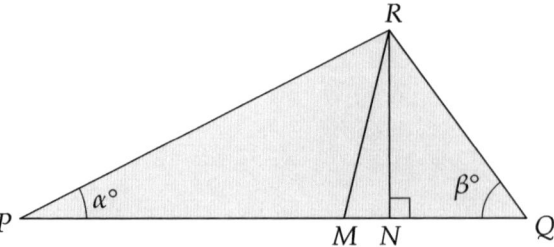

     What is the size of $\angle NRM$?

22   **6.** The diagram shows a regular dodecagon.
     What is the size of the marked angle?

     [*A* dodecagon *is a polygon with twelve sides.*

     In a regular polygon, *all the sides are equal and all
     the interior angles are equal.*]

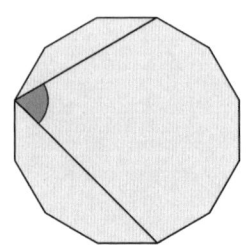

2 **7.** In triangle *PQR* the point *N* lies on the side *PR*, and the point *M* lies on the side *PQ*. The segment *QN* bisects the angle *PQR*, and *MN* is parallel to *QR*.

Which of the following five statements follows logically from this given information?

$MN = \frac{1}{2}QR$ $MN = MQ$ $MN = NP$ $MN = RN$
the other four statements may all be false

3 **8.** In triangle *PQR*, *PS* = 2; *SR* = 1; ∠*QRP* = 45°; *T* is the foot of the perpendicular from *P* to *QS* and ∠*TSP* = 60°.

What is the size of ∠*RPQ*?

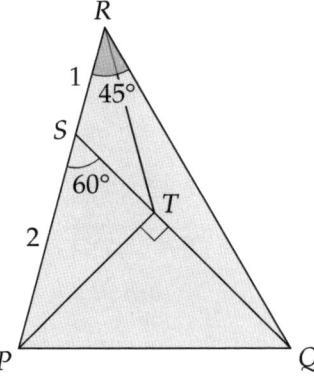

3 **9.** What is the greatest number of the following five properties that a single heptagon can possibly possess?

Its interior angles add up to 900°.

All its sides are equal.

It has exactly four acute interior angles.

It has exactly one line of symmetry.

It has no obtuse interior angles.

[A heptagon *is a polygon with seven sides.]*

3 **10.** *PQRSTUVWX* is a regular polygon with nine sides. What is the size of angle *TPU*?

[A polygon with nine sides is called a nonagon *or* enneagon.*]*

25 **11.** The regular hexagon in the diagram has been divided into four trapezia and one hexagon.   Each of the five sections has the same perimeter length.

What is the ratio of the lengths p, q and r?

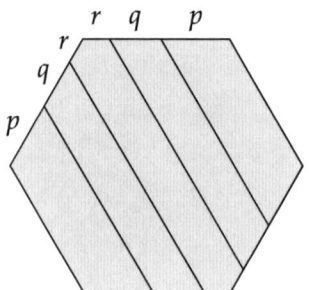

More triangles

Exercise 33

7 **1.** The diagram shows a triangle PQR with
   $PQ = QR = 9$ cm and $RP = 6$ cm.
   The point S lies on QR so that $RP = PS$.
   What is the length of SR?

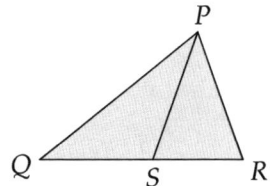

7 **2.** Which of the following four triangles is right-angled?.

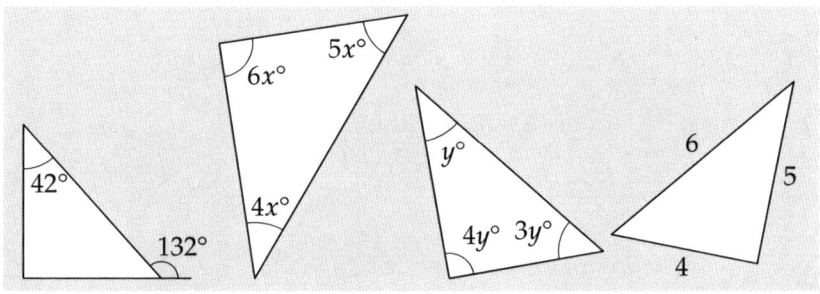

8 **3.** The sum of the areas of the squares on the sides of a right-angled
   isosceles triangle is $72\,\text{cm}^2$.
   What is the area of the triangle?

18   **4.** In the triangle PQR, there is a right
      angle at Q and angle RPQ is 60°. The
      bisector of the angle RPQ meets QR
      at S, as shown.

      What is the ratio $QS : SR$?

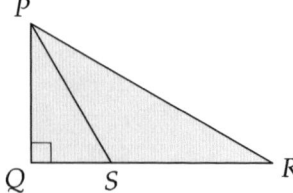

19   **5.**   Harrogate is 23 km due north of Leeds.

                      York is 30 km due east of Harrogate.

                      Doncaster is 48 km due south of York.

                      Manchester is 70 km due west of Doncaster.

      To the nearest kilometre, how far is it from Leeds to Manchester, as
      the crow flies?

20   **6.** Jack's teacher asked him to draw a triangle of area $7\,\text{cm}^2$. Two sides
      are to be of length 6 cm and 8 cm.

      How many possibilities are there for the length of the third side of the
      triangle?

20   **7.** In the right-angled triangle PQR,
      $QS = 8$, $PS = 10$ and $PS = SR$.

      What is the area of triangle PSR?

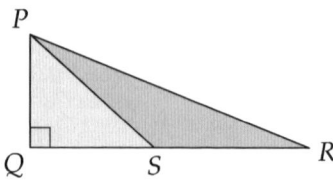

21   **8.** The parallelogram $PQRS$ is formed
      by joining together four equilateral
      triangles with sides of length 1, as
      shown.

      What is the length of the diagonal QS?

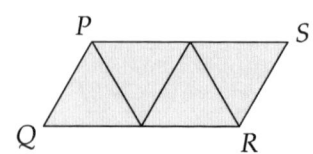

9. Two circles with radii 1 cm and 4 cm touch, as shown. The point *P* is on the smaller circle, *Q* is on the larger circle and *PQ* is a tangent to both circles.

   What is the length of *PQ*?

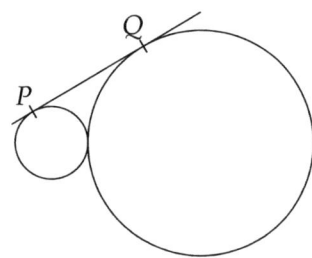

10. The diagrams show squares placed inside two identical semicircles. In the lower diagram the two squares are identical.

    What is the ratio of the areas of the two shaded regions?

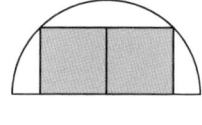

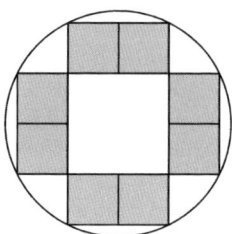

11. The diagram shows a pattern of eight equal shaded squares inside a circle of area π.

    What is the area of the shaded region?

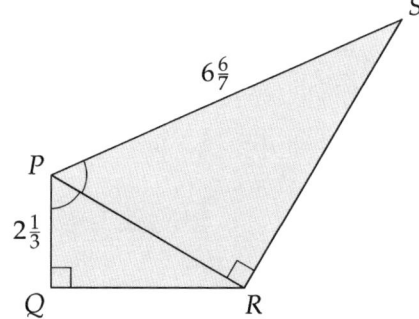

12. In the diagram, $PQ = 2\frac{1}{3}$, $SP = 6\frac{6}{7}$, and $\angle RPQ$ is equal to $\angle SPR$.

    How long is *PR*?

24   **13.** The diagram has rotational symmetry
of order four about S. Angle PQR is
equal to $15°$ and the area of $PQTU$ is
equal to 24.

What is the length of RS?

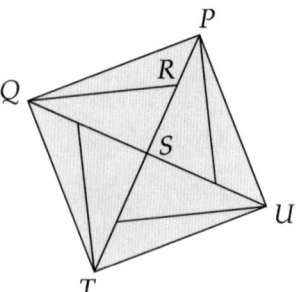

24   **14.** The diagram shows a square of area x
inscribed in a semicircle, and a larger square
of area y inscribed in a circle.

What is the ratio $x : y$?

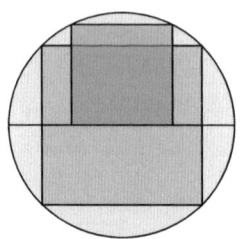

24   **15.** The diagram shows a $1 \times x$ rectangular plank
that fits neatly inside a 10×10 square frame.

What is the value of x?

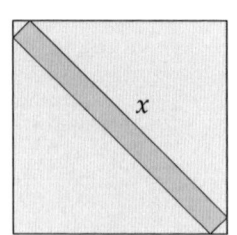

25   **16.** The diagram shows two concentric circles
with radii 1 and 2, together with a shaded
octagon, all of whose sides have equal length.

What is the length of the perimeter of the
octagon?

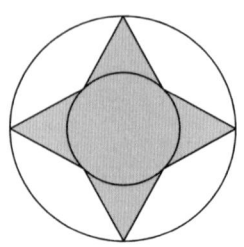

25 **17.** A garden has the shape of a right-angled triangle with sides of length 30, 40 and 50. A straight fence goes from the corner with the right angle to a point on the opposite side, dividing the garden into two sections which have the same perimeter length.

How long is the fence?

25 **18.** A rectangular sheet of paper with sides of length 1 and $\sqrt{2}$ has been folded once as shown, so that one corner just meets the opposite long side.

What is the value of d?

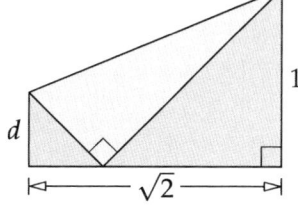

More areas

Exercise 34

1. The point O is the centre of a circle of radius 1, OP and OR are radii, and $OPQR$ is a square.

What is the area of the shaded region?

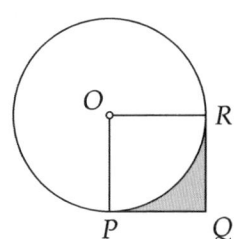

2. Three-quarters of the area of the rectangle has been shaded.

What is the value of x?

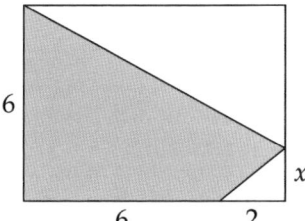

18   **3.** What fraction of the rectangle *PQRS* is shaded?

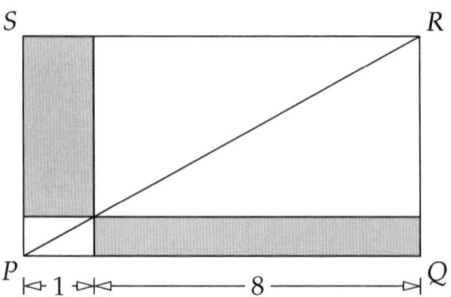

19   **4.** The diagram shows four smaller squares in
      the corners of a large square. The smaller
      squares have sides of length 1 cm, 2 cm, 3 cm
      and 4 cm (in anticlockwise order) and the
      sides of the large square have length 11 cm.

      What is the area of the shaded quadrilateral?

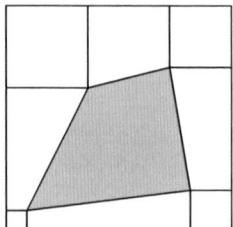

19   **5.** In terms of *a*, *b* and *c*, what is the area
      of the pentagon shown?

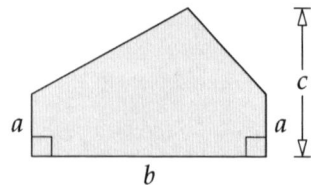

20   **6.** What is the area of this
      quadrilateral?

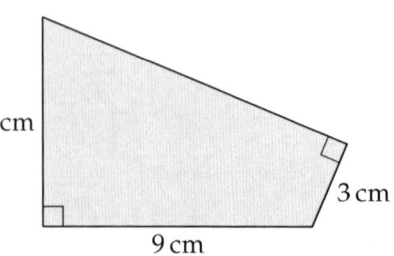

7. A rectangle is placed obliquely on top of an identical rectangle, as shown. The area of the overlapping region (shaded more darkly) is one eighth of the total shaded area.

What fraction of the area of one rectangle is the area of the overlapping region?

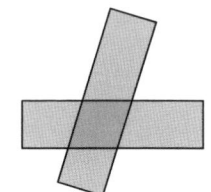

8. The square *PQRS* has an area of 196. It contains two overlapping squares; the larger of these squares has an area 4 times that of the smaller and the area of their overlap is 1.

What is the total area of the shaded regions?

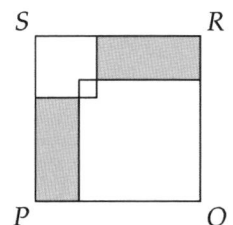

9. In square *RSTU*, a quarter-circle with centre *S* is drawn from *T* to *R*. The point *P* lies on this arc, at a distance 1 from *TU* and 8 from *RU*, as shown.

What is the length of the side of square *RSTU*?

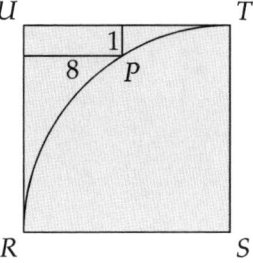

10. A point is marked one quarter of the way along each side of a triangle, as shown.

What fraction of the area of the triangle is shaded?

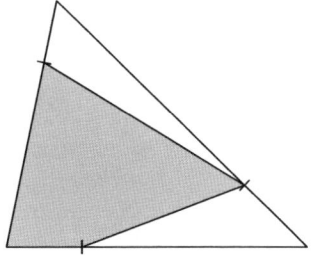

25   **11.** The diagram shows a square, a diagonal and a
     line joining a vertex to the midpoint of a side.

     What is the ratio of area P to area Q?

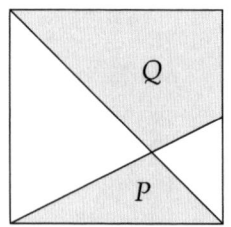

25   **12.** Two squares, each of side-length $1 + \sqrt{2}$, overlap. The overlapping
     region is a regular octagon.

     What is the area of the octagon?

25   **13.** The large circles in the following five diagrams have the same radius.
     Which shaded area is the greatest?

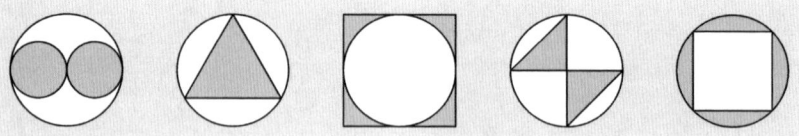

25   **14.** A square is inscribed in a 3-4-5 right-angled
     triangle as shown.

     What fraction of the area of the triangle does
     the square occupy?

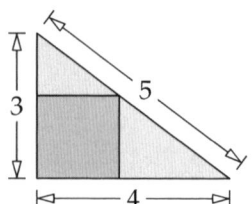

More counting

Exercise 35

1. Pat is asked to draw two different straight lines parallel to the x-axis, three different straight lines parallel to the y-axis, and four different straight lines parallel to the line $y = x$.

   What is the smallest possible total number of crossing points among the nine lines that Pat draws?

2. A voucher code is made up of four characters.
   - (i) The first is a letter: V, X or P.
   - (ii) The second and third are different digits.
   - (iii) The fourth is the units digit of the sum of the second and third digits.

   How many different voucher codes like this are there?

3. Shahbaz thinks of a positive integer n such that the difference between $\sqrt{n}$ and 7 is less than 1.

   How many different possibilities are there for n?

4. In how many different ways can seven different numbers be chosen from the numbers 1 to 9 inclusive so that the seven numbers have a total which is a multiple of three?

21 **5.** A square maze has nine rooms with gaps in the walls between them. Once a person has travelled through a gap in the wall it closes behind them.

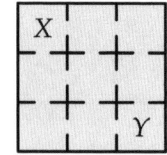

How many different ways can someone travel through the maze from X to Y?

21 **6.** There are lots of ways of choosing three dots from this 4×4 array.

How many triples of points are there where all three lie on a straight line (not necessarily equally spaced)?

22 **7.** A square is divided into eight congruent right-angled isosceles triangles, as shown. Two of these triangles are selected at random and shaded black.

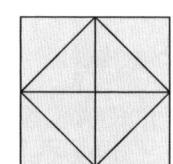

What is the probability that the resulting figure has at least one line of symmetry?

22 **8.** An 8×8 chessboard is placed so that a black square is in the top left-hand corner. Starting in the top left cell and working along each row in turn, coloured counters are placed, one on each cell, following the sequence black, white, red, black, white, red and so on. When the right-hand end of each row is reached, the pattern continues, starting at the left-hand end of the row beneath, until there is one counter on every cell.

In the final arrangement, what fraction of the counters are on cells of the same colour as themselves?

22 **9.** One hundred and twenty students take an examination which is marked out of 100 (with no fractional marks). No three students are awarded the same mark.

What is the smallest possible number of pairs of students who are awarded the same mark?

23 **10.** There are 120 different ways of arranging the letters, U, K, I, M and C. All of these arrangements are listed in dictionary order, starting with CIKMU.

Which position in the list does UKIMC occupy?

24 **11.** How many four-digit integers (from 1000 to 9999) have at least one digit repeated?

24 **12.** A new taxi firm needs a memorable phone number. They want a number which has a maximum of two different digits. Their phone number must start with the digit 3 and be six digits long.

How many such numbers are possible?

24 **13.** In the diagram on the right, how many squares, of any size, are there whose entries add up to an even total?

1	2	3	4	5
6	7	8	9	10
11	12	13	14	15
16	17	18	19	20
21	22	23	24	25

25 **14.** How many different solutions are there to the following crossnumber?

ACROSS
1. Prime
3. Square
5. Prime

DOWN
1. Prime
2. Square
4. Square

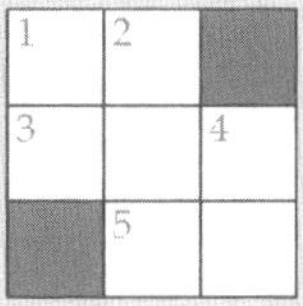

[A crossnumber is like a crossword, except that all the answers are numbers instead of words, with one digit in each cell, and no answer starts with the digit zero.]

More reasoning

Exercise 36

.6 **1.** The diagram shows an L-shape divided into 1×1 squares. Gwyn cuts the shape along some of the lines shown to make two pieces, neither of which is a rectangle (or a square). She then uses the pieces to form a rectangle.

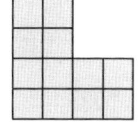

What is the difference between the areas of the two pieces?

.8 **2.** One of the digits 1 to 9 is put in each unshaded cell so that no digit is repeated and the totals of the entries in the rows and columns are as shown.

What number goes in the starred cell?

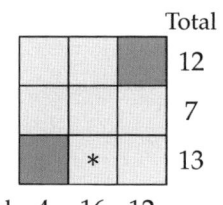

20   **3.** A competition has 25 questions and is marked as follows.

> Five marks are awarded for each correct answer to Questions 1–15.
> Six marks are awarded for each correct answer to Questions 16–25.
> Each incorrect answer to Questions 16–20 loses 1 mark.
> Each incorrect answer to Questions 21–25 loses 2 marks.

Which of the following five scores is it impossible to achieve?

126   127   128   129   130

22   **4.** In a particular group of people, some always tell the truth, the rest always lie. There are 2016 in the group.

One day, the group is sitting in a circle. Each person in the group says, "Both the person on my left and the person on my right are liars."

What is the difference between the largest and smallest number of people who could be telling the truth?

23   **5.** Four brothers are discussing the order in which they were born. Two are lying and two are telling the truth.

> Alfred: "Bernard is the youngest."
> Horatio: "Bernard is the oldest and I am the youngest."
> Inigo: "I was born last."
> Bernard: "I'm neither the youngest nor the oldest."

Which two are telling the truth?

23   **6.** In the star shown the sum of the four numbers in any "line" is the same for each of the five "lines". The five missing numbers are 9, 10, 11, 12 and 13.

Which number is represented by K?

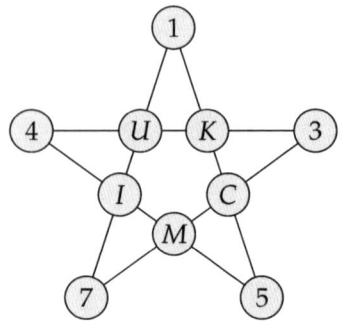

23 **7.** In this unusual game of noughts and crosses the first player to form a line of three Os or three Xs *loses*. It is X's turn.

Where should she place her cross to make sure that she does not lose?

24 **8.** What is the largest number of the following five statements that can be true at the same time?

$$0 < x^2 < 1 \qquad x^2 > 1 \qquad -1 < x < 0 \qquad 0 < x < 1$$
$$0 < x - x^2 < 1$$

24 **9.** A 4×4 'antimagic square' is an arrangement of the numbers 1 to 16 inclusive in a square, so that the totals of each of the four rows and four columns and two main diagonals are ten consecutive numbers in some order. The diagram shows an incomplete antimagic square.

4	5	7	14
6	13	3	*
11	12	9	
10			

When it is completed, what number will replace the asterisk?

24 **10.** The Queen of Hearts has lost her tarts! She is sure that those knaves who have not eaten the tarts will tell her the truth and that the guilty knaves will tell lies. When questioned, the five knaves declare

> Knave 1: "One of us ate them."
>
> Knave 2: "Two of us ate them."
>
> Knave 3: "Three of us ate them."
>
> Knave 4: "Four of us ate them."
>
> Knave 5: "Five of us ate them."

How many of the knaves were honest?

25 **11.** One coin among N identical-looking coins is a fake and is slightly heavier than the others, which all have the same weight. To compare two groups of coins you are allowed to use a set of scales with two pans which balance exactly when the weight in each pan is the same.

What is the largest value of N for which the fake coin can be identified using a maximum of two such comparisons?

More dissections

Exercise 37

18 **1.** The triangle T has sides of length 6, 5, 5. The triangle U has sides of length 8, 5, 5.

What is the ratio area T : area U?

18 **2.** The diagram contains six equilateral triangles with sides of length 2 and a regular hexagon with sides of length 1.

What fraction of the whole shape is shaded?

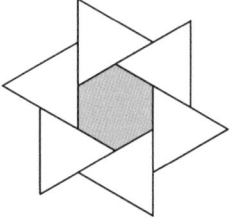

21 **3.** The diagram shows a regular octagon.

What is the ratio of the area of the shaded trapezium to the area of the whole octagon?

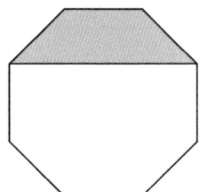

21   **4.** A regular octagon is placed inside a square, as shown. The shaded square connects the midpoints of four sides of the octagon.

What fraction of the outer square is shaded?

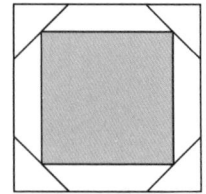

21   **5.** A piece of thin card in the shape of an equilateral triangle with sides of length 3 and a circular piece of thin card with radius 1 are glued together so that their centres coincide.

How long is the outer perimeter of the resulting two-dimensional shape?

22   **6.** The diagram shows a shaded region inside a large square. The shaded region is divided into small squares.

What fraction of the area of the large square is shaded?

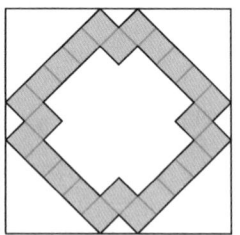

23   **7.** The diagram shows part of a tiling pattern which is made from two types of individual tiles: 8×6 rectangular grey tiles and square black tiles.

The pattern is extended to cover an infinite plane.

What fraction of the plane is coloured black?

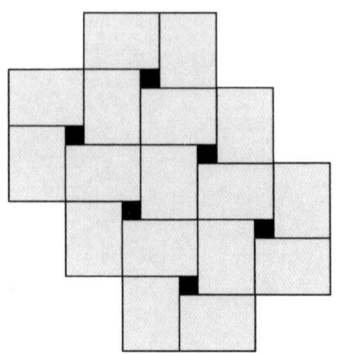

24 **8.** The length of each side of this regular octagon is 2 cm.

What is the difference between the area of the shaded region and the area of the unshaded region?

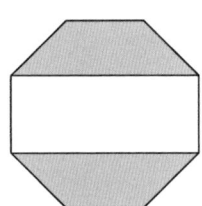

25 **9.** The tiling pattern shown uses two types of tile—light grey regular hexagons and dark grey equilateral triangles—with the length of each side of the equilateral triangles equal to half the length of each side of the hexagons.

A large number of these tiles is used to cover a floor.

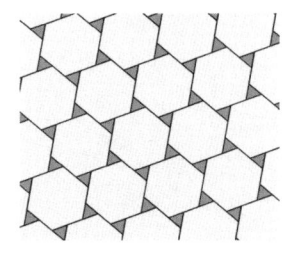

Which of the following five fractions is closest in value to the fraction of the floor that is shaded dark grey?

$$\frac{1}{8} \qquad \frac{1}{10} \qquad \frac{1}{12} \qquad \frac{1}{13} \qquad \frac{1}{16}$$

25 **10.** The diagram shows a ceramic design by the Catalan architect Antoni Gaudi. It is formed by drawing eight lines connecting points which divide the sides of the outer regular octagon into three equal parts, as shown.

What fraction of the octagon is shaded?

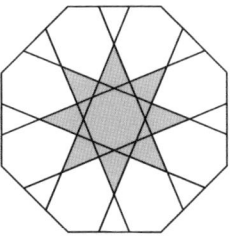

25 **11.** The diagram shows a square with two lines from a corner to the middle of an opposite side. The rectangle fits exactly inside these two lines and the square itself.

What fraction of the square is occupied by the shaded rectangle?

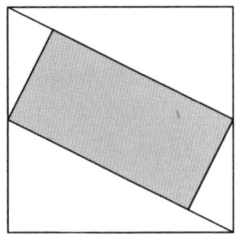

Part III

Remarks and answers

> What is best in mathematics deserves
> not merely to be learnt as a task, but to
> be assimilated as a part of daily thought,
> and brought again and again before the
> mind with ever-renewed encouragement.

Bertrand Russell
The Study of Mathematics

The remarks are intended to help you to arrive at the answer, possibly using a different approach to any you may have in mind.

An answer is given for every problem in the book.

Exercise 1

1. ANSWER: 4086

2. The expression is equal to $0.8 + 0.026$.

ANSWER: 0.826

3. The expression is equal to $4.5 \times (5.5 + 4.5)$.

ANSWER: 45

4. The expression is equal to $10 + 10 \times 10 \times 20$.

ANSWER: 2010

5. The expression is equal to $1 + 8 + 20$.

ANSWER: 29

6. 4.004 is less than 4.04.

ANSWER: 4.044

7. ANSWER: 1998

8. The difference between -3 and 8 is equal to $8 - (-3)$, which is equal to 11.

ANSWER: $-3, 8$

9. The value is equal to $5 \times 0.3 + 2.1$.

ANSWER: 3.6

10. The value of 0.3×7 is equal to 2.1.

ANSWER: 0.09×30

11. 37 373 pairs with 61 392.

ANSWER: 70 082

12. The value of $634\,566 \div 2786$ is equal to 231.

ANSWER: 23 100

13. She needs to divide by 10.

ANSWER: ɪ˙0 ʎq ʎๅdๅๅnɯ

14. Three cans are enough for fifteen children.

ANSWER: xıs

15. The expression in brackets is equal to $1234 \times (10 + 0.01)$.

ANSWER: ๅ0˙0ๅ

16. Multiplying the answer by 0.2 gives 30.

ANSWER: 0Ṣๅ

Exercise 2

1. 3333 is equal to 3 × 1111, and is therefore not prime.

ANSWER: one

2. Both (i) and (ii) are true for the positive integers 1, 9, 15 and 21, but (iii) is false.

ANSWER: 25

3. The integers are 1, 4, 6, 8 and 9.

ANSWER: 28

4. The expression is equal to 40.

ANSWER: 5

5. $1^2 + 2^2$ is equal to 5, and is therefore prime.

ANSWER: one

6. ☛ *A positive integer is a multiple of 3 when its digits add up to a multiple of 3, and not otherwise.*

The sum of the digits of 234 567 is 27, thus 234 567 is a multiple of 3 and is therefore *not* prime.

ANSWER: 23 456 789

7. $2^2 - 1$ is equal to 3, and is therefore prime.

ANSWER: $2^6 - 1$

8. Though 45 is equal to 3 × 3 × 5 and is thus the product of three primes, 45 is not the product of three *distinct* primes.

ANSWER: 105

9. Neither prime is 2 or 3.

ANSWER: 16

Exercise 3

1. The value is equal to $\dfrac{1}{4} \times \dfrac{3}{4}$.

ANSWER: $\dfrac{3}{16}$

2. The two fractions are $\dfrac{6}{24}$ and $\dfrac{4}{24}$.

ANSWER: $\dfrac{5}{24}$

3. The value is equal to $3 \times \frac{2}{1}$.

ANSWER: 6

4. ♡ lies between 3×5 and 4×5.

ANSWER: 19

5. The original number is 4×24.

ANSWER: 32

6. $\dfrac{7}{8}$ is $\dfrac{1}{8}$ less than 1.

ANSWER: $\dfrac{11}{10}$

7. The value is equal to $\dfrac{2}{4} \times \dfrac{5}{15} + \dfrac{5}{15} \times \dfrac{1}{4} + \dfrac{1}{4} \times \dfrac{3}{15}$.

ANSWER: three-tenths

8. The value is equal to $\frac{3}{4} \div \frac{3}{4}$.

ANSWER: 1

9. The two fractions are $\dfrac{4}{16}$ and $\dfrac{2}{16}$.

ANSWER: $\dfrac{3}{16}$

10. Four of the five fractions are less than $\frac{1}{2}$; one is greater.

Answer: $\dfrac{11}{9}$

11. Victoria eats $\dfrac{8}{12} - \dfrac{3}{12}$ of the cake.

Answer: five-twelfths

12. $\dfrac{8}{56}$ is equal to $\dfrac{1}{7}$.

Answer: $\dfrac{1+4}{7+4}$

13. One half of $\dfrac{1}{25}$ is equal to $\dfrac{1}{50}$.

Answer: one sixth of $\dfrac{1}{5}$

14. ☛ *A positive integer is a multiple of 9 when its digits add up to a multiple of 9, and not otherwise.*

☛ *When the highest common factor of the positive integers p and q is equal to 1, an integer that is a multiple of p and a multiple of q is also a multiple of pq.*

The digits of 594 add up to 18, thus 594 is a multiple of 9; it is also a multiple of 2, and therefore is a multiple of 18.

Answer: $\dfrac{873}{8+7+3}$

15. The value is equal to $67 + 67$.

Answer: 134

16. The difference between the number of sheep and the number of goats is one third of the animals.

Answer: 36

17. The two fractions are $\dfrac{12}{15}$ and $\dfrac{10}{15}$.

ANSWER: $\dfrac{11}{15}$

18. $\dfrac{2+3}{4+6}$ is equal to $\dfrac{1}{2}$.

ANSWER: $\dfrac{2 \times 3}{4 \times 6}$

Exercise 4

1. ☞ *A positive integer is a multiple of 3 when its digits add up to a multiple of 3, and not otherwise.*

   No odd number is divisible by 6.
   Also, 99 998 is not divisible by 3, so 99 998 is not divisible by 6.

   ANSWER: one million minus four

2. The units digit of the answer is the units digit of 7×9.

   ANSWER: 3

3. The three integers are 1, 2 and 4.

   ANSWER: 8

4. The number is equal to $10\,000\,000 + 100\,000 + 1$.

   ANSWER: 10 100 001

5. The division may be written as the multiplication 'pq' $= r \times s$.

   ANSWER: 8

6. Let the numbers of stools and chairs be s and c respectively, so that $3s + 4c = 17$.
   Hence s is an odd positive integer, at most 5.

   ANSWER: 3

7. The answer is the highest common multiple of 5 and 7.

   ANSWER: 35

8. ☞ *The product of any integer and an even integer is even.*
   ☞ *The product of two odd integers is odd.*
   ☞ *The sum of two even integers is even.*
   ☞ *The sum of two odd integers is even.*
   ☞ *The sum of an even integer and an odd integer is odd.*

   The expression $3m + 4n$ is even.

   ANSWER: $(m + 3n)^2$

9. The tens digits are 9 and 8, and the units digits are 7 and 6.

   ANSWER: 183

10. Martha has $5 \times 4 \times 3$ grandchildren.

    ANSWER: 85

11. The total cost of the apples is $9 \times 20p$.

    ANSWER: 20p

12. The next palindromic number is 25052.

    ANSWER: 110

13. The nine integers are n to $n + 8$ for some n.

    ANSWER: 8

14. Let the original integer be n; then the answer to the calculation is

$$4 \times (2n + 3) - 5 - n.$$

    ANSWER: 7

15. Four cups and five saucers were on the tray.

    ANSWER: three

Exercise 5

1. There are 24 × 7 hours in a week in which the clocks do not change.

   ANSWER: 168

2. ANSWER: 2024

3. The machine cracks open $\dfrac{180\,000}{60 \times 60}$ eggs per second.

   ANSWER: 50

4. There are 31 × 24 hours in one of the longest months.

   ANSWER: 744

5. The number of calories per day is equal to 7000 ÷ 365.

   ANSWER: 20

6. The cycle of bells repeats every four hours, and the bell is struck 36 times in each cycle.

   ANSWER: 216

7. Ten minutes less than 25 hours elapse.

   ANSWER: 1490

8. One banana weighs around 160 grams; 250 kg is 250 000 grams.

   ANSWER: 4 or 5 a day

9. ☞ *When divided by 3, a positive integer and the sum of its digits have the same remainder.*

   The Festival is held in Worcester in years that have a remainder of 1 when divided by 3.

   ANSWER: 2054

10. The next time that all the digits 0, 1, 2, 3 appear is 03:12.

    ANSWER: 41

11. October started on a Saturday that year.

ANSWER: Monday

12. Sydney's flight arrived at 8:30 am on Wednesday *London* time.

ANSWER: 9:30 pm on Tuesday

13. There are 4 × 24 quarters of an hour in a 24-hour day. School takes up 27 of these.

ANSWER: $\dfrac{9}{32}$

14. On average he added 6 000 000 ÷ 5 rubber bands per year.

ANSWER: 3300

15. There are 60 × 60 × 24 seconds in a day, which is roughly 100 000.

ANSWER: 3×10^7

16. There are $\dfrac{1000}{4} - \dfrac{1000}{100}$ potential leap years between 2001 and 3001, except that some of these years are multiples of 400.

ANSWER: 242

17. ☛ *A year with 365 days advances the days of the week for the following year by one, and a leap year advances them by two.*

ANSWER: 3

18. The next time that the digits are all the same is 11:11.

ANSWER: 316

19. The fine is just over £30 per year.

ANSWER: £2200

20. The number of minutes in a week is 60 × 24 × 7.

ANSWER: 500 000

Exercise 6

1. The number of chocolate ice-creams sold was $1\frac{1}{2}$ times the number of strawberry ice-creams sold.

ANSWER: 06

2. The diagram alongside shows how to make the first shape.

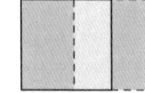

ANSWER:

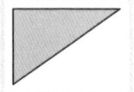

3. Any straight line passing through the centre of the square cuts it in half (see the diagram alongside).

ANSWER: infinitely many

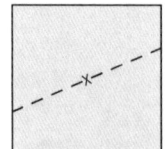

4. The outcome on the next roll does not depend on what has happened before.

ANSWER: $\frac{9}{1}$

5. The dashed lines in the diagram alongside are the only possible lines of symmetry.

ANSWER: 3

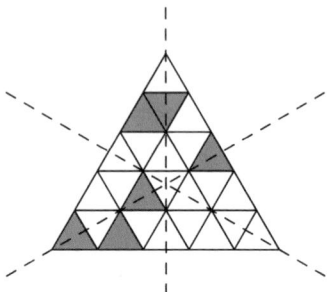

6. The positions of the villages are shown in the diagram alongside.

ANSWER: north-east

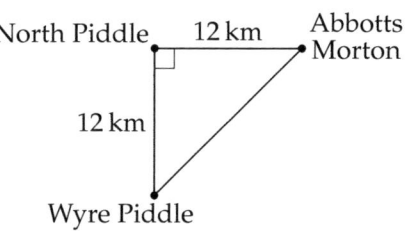

7. The halfway point is $200 - 3\frac{1}{2}$ miles from Edinburgh,

ANSWER: 393

8. The sum would normally be written $162 + 257$.

ANSWER: 580

9. The letter-product of 6 is 6×3.

ANSWER: 8

10. Points equidistant from Q and R lie on the dashed line in the diagram alongside, and points exactly 6 cm from P lie on the dashed circle.

ANSWER: 2

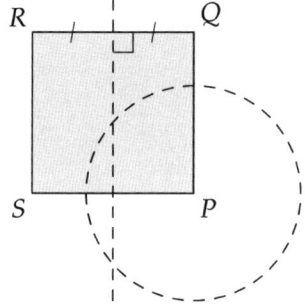

11. The angle does not change when it is magnified.

ANSWER: $2.5°$

12. ☞ *Pythagoras' Theorem.*

☞ *For positive a and b: a is less than b when a² is less than b², and not otherwise.*

Let the length of the short sides of the isosceles right-angled triangle be 1. Therefore the length of the hypotenuse is $\sqrt{2}$.

But $2\sqrt{2}$ is less than 3 because $\left(2\sqrt{2}\right)^2$ is less than 9.

Answer:

13. Two touching coins can never be removed. Removing one of the hatched coins in the diagram alongside allows at most two other coins to be removed.

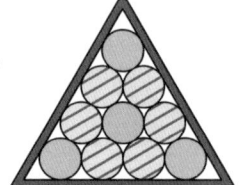

Answer: ᔭ

14. When Meg stands on her brother's shoulders she is 2.5 times as tall as normal.

Answer: ɯ ϛ˙ㄥ

15. Since 2004 is a multiple of 12, the section between 2007 and 2011 looks like the section between 3 and 7.

Answer:

16. The sum is $5 + 4 = 9$.

Answer: uǝǝʇuǝʌǝs

17. The middle diagonal has one more white dot than black dot.

Answer: ₂Ɫ00Ⅰ + ₂000Ⅰ

18. ANSWER: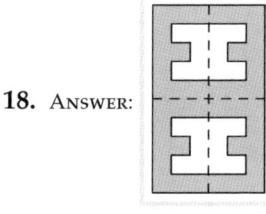

19. ANSWER: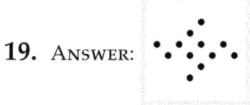

20. Suppose that Jack originally had the cards in the pile on the table in the order (top to bottom) *abcde*. Then Jack gave me the cards in the order *acedb*.

ANSWER: Ace, 5, 2, 4, 3

Exercise 7

1. ☞ *For any power p:* $1^p = 1$.

   ☞ *For any positive a, b and any positive integer n:* a^n *is less than* b^n *when a is less than b.*

   $1^9 \times 9^9$ is equal to 9^9, which is less than 199^9.

   ANSWER: 1999

2. $16 = 4^2$ and thus 16 is a square.

   ANSWER: 80

3. $\left(1^2 + 1\right)\left(10^2 + 1\right)$ is a multiple of 2, whereas 2005 is not.

   ANSWER: $\left(2^2 + 1\right)\left(20^2 + 1\right)$

4. $1^6 = \left(1^3\right)^2$ and thus 1^6 is a square.

   ANSWER: 25

5. ☞ *For any a and b:* $(a + b)^2 = a^2 + 2ab + b^2$.

   The expression is equal to $(2000 + 3)^2$.

   ANSWER: 4012009

6. The value of $1^0 - 0^1$ is $1 - 0$.

   ANSWER: $5^4 - 4^5$

7. The units digit of the two-digit square is 9 or 1.

   ANSWER: 2

8. The value of 2^{10} is 1024.

   ANSWER: 924

9. ☞ *For any p, q and positive a:* $a^p \times a^q = a^{p+q}$.

   Half of 2^{20} is equal to $2^{-1} \times 2^{20}$.

   ANSWER: 219

10. 10 is halfway between 4 and 4^2.

An integer halfway between a positive integer n and n^2 is equal to $\frac{1}{2} \times (n + n^2)$, and the equation $n^2 + n = 60$ has no integer solutions.

ANSWER: 0Ɛ

11. ☞ *The product of two even integers is a multiple of 4.*

The units digit of any positive integer power of 66 is 6. Therefore the answer is 3 or 8. But 66^{66} is a multiple of 4, thus half of it is even.

ANSWER: 8

12. ☞ *For positive a and b: a is less than b when a^2 is less than b^2, and not otherwise.*

$\left(3\sqrt{11}\right)^2 = 9 \times 11$, which is less than 10^2.

ANSWER: Ɛ

Exercise 8

1. The rings indicated in the diagram alongside need to be separated.

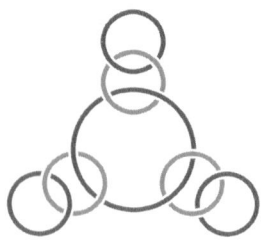

ANSWER: three

2. Given any one pupil, there are at least 24 others with a different birthday.

ANSWER: 9

3. The only possible imperial units of length that can be spelled out simultaneously are 'MILE' and 'YARD'.

ANSWER:
C
Y

4. The first and third statements are true. It is not difficult to find examples showing that the second and fourth statements are false, such as 18 and $2 + 2 = 4$.

ANSWER: two

5. From the diagonal, the 'magic' total is 58. It is now possible to find the unknown numbers in the first column.

ANSWER: 21

6. Kevin belongs in K and L, but not in J, nor in M.

ANSWER: E

7. The Queen could say the first statement: if she is lying that day, then the statement is false (so she lied the day before too); if she is telling the truth, then the statement is true (so she told the truth the day before too).

ANSWER: Today, I lie.

8. Since the sum of the digits in each ring is 11, the 2, 3 and 4 can be placed immediately. Let the three other missing numbers (1, 6 and 7) be a, b and c, as in the diagram below.

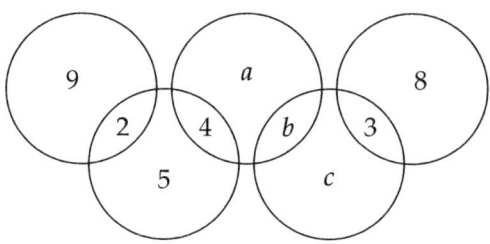

Because $a + 4$ is equal to $c + 3$, it follows that c is one more than a.

ANSWER: 9

9. Omitting each word in turn, the total number of letters in the other four is 38, 35, 33, 35 or 35.

ANSWER: thirty-three

10. Exactly one of Jenny and Willie was telling the truth, and exactly one of Sam and Mrs Scrubitt was.

ANSWER: 2

11. If Tweedledee's statement is false, then Tweedledum's is true, and vice versa.

Hence the statements made by Alice and the White Rabbit are both false.

ANSWER: 1

12. The *only* way to be certain she wins, whatever White plays, is for Black to form a line of three counters with an empty cell at both ends.

ANSWER: in the fourth cell from the left on the bottom row

13. Let the missing numbers be a, b, c, d, e, f and g, as shown in the diagram alongside, and let the total of the numbers in each of the lines of three circles be T.

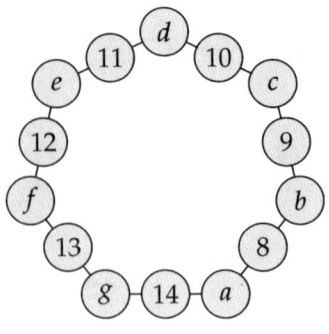

Then $a + b + c + d + e + f = 28$, and

$$7T = 77 + 2(a + b + c + d + e + f).$$

It is possible to fill in all the circles.

ANSWER: 61

14. The only guesses that could have been out by 70 g and 90 g were 5060 g and 5040 g.

ANSWER: 5130 g

15. Suppose that there are b boys in the family, then there are $b + 1$ girls, and each boy has $b + 1$ sisters and $b - 1$ brothers.
Hence $b + 1 = \frac{3}{2} \times (b - 1)$.

ANSWER: 11

Exercise 9

1. ☞ *Vertically opposite angles are equal.*

Any arrangement of the points may be obtained from the diagram alongside by a rotation about *O*.

Both *TOU* and *VOW* are straight lines.

ANSWER: 18°

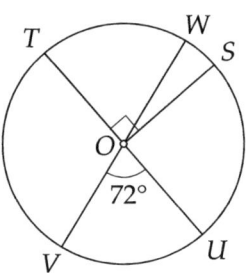

2. ☞ *The angles of a triangle add up to 180°.*

Let the angles be $2k°$, $3k°$ and $5k°$.

ANSWER: 54°

3. ☞ *Corresponding angles on parallel lines are equal.*

☞ *An exterior angle of a triangle is equal to the sum of the interior opposite angles.*

In the diagram alongside, the marked angle is equal to $x°$ and the angle $y°$ is an exterior angle of the shaded triangle.

ANSWER: $y = 2x$

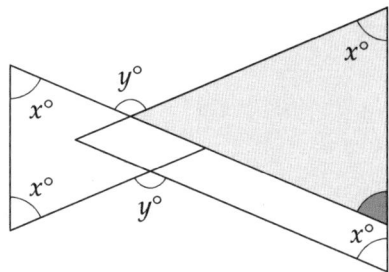

4. ☞ *The exterior angles of a triangle add up to 360°.*

ANSWER: 720

5. ☞ *Alternate angles on parallel lines are equal.*

   ☞ *Corresponding angles on parallel lines are equal.*

   ☞ *An exterior angle of a triangle is equal to the sum of the interior opposite angles.*

In the diagram alongside, the angle $x°$ is an exterior angle of the shaded triangle.

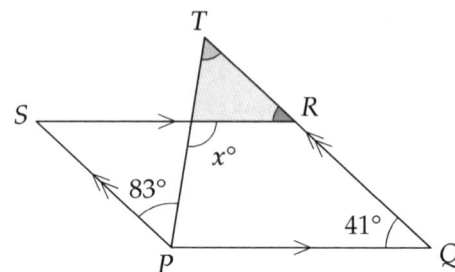

ANSWER: ⏉ΖІ

6. ☞ *The base angles of an isosceles triangle are equal.*

   ☞ *An exterior angle of a triangle is equal to the sum of the interior opposite angles.*

In the diagram below, the marked angle is an exterior angle of the shaded triangle.

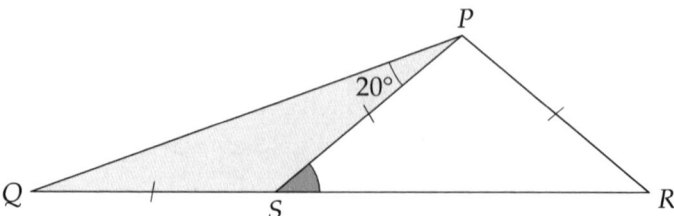

ANSWER: ₒ0⏉

7. ☞ *Angles at a point add up to 360°.*

   ☞ *The interior angles of a polygon with n sides add up to $(n-2) \times 180°$.*

The missing interior angles of the hexagon are $285°$ and $360° - x°$ and so

$$285 + 27 + 24 + 360 - x + 23 + 26 = 720.$$

ANSWER: ϚΖ

8. ☛ *Angles on a straight line add up to 180°.*

When you travel once around the figure, along the five arrows in the diagram alongside, you make one complete turn. Hence

$$180 - x + 100 + 110 = 360.$$

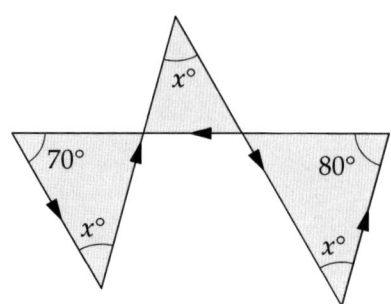

ANSWER: 0Ɛ

9. ☛ *Vertically opposite angles are equal.*

☛ *The base angles of an isosceles triangle are equal.*

☛ *Alternate angles on parallel lines are equal.*

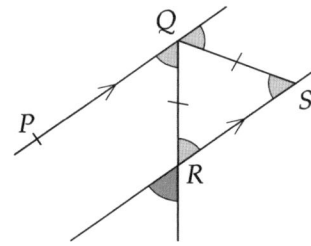

ANSWER: ɹnoɟ

10. ☛ *The base angles of an isosceles triangle are equal.*

☛ *The angles of a triangle add up to 180°.*

☛ *An exterior angle of a triangle is equal to the sum of the interior opposite angles.*

The marked angle in the diagram alongside is equal to 72°.

ANSWER: 80Ɩ

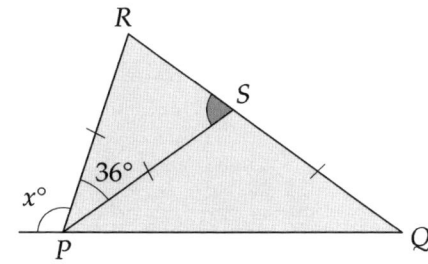

11. ☞ *An exterior angle of a triangle is equal to the sum of the interior opposite angles.*

☞ *Angles on a straight line add up to 180°.*

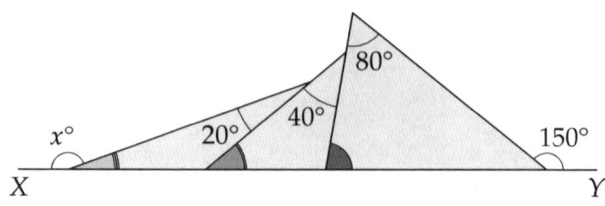

ANSWER: 0ㄥᴉ

12. ☞ *The interior angles of a quadrilateral add up to 360°.*

The value of x is 30.

ANSWER: unızǝdɐɹʇ

13. ☞ *The opposite angles of a parallelogram are equal.*

☞ *The base angles of an isosceles triangle are equal.*

☞ *The angles of a triangle add up to 180°.*

☞ *Allied angles on parallel lines add up to 180°.*

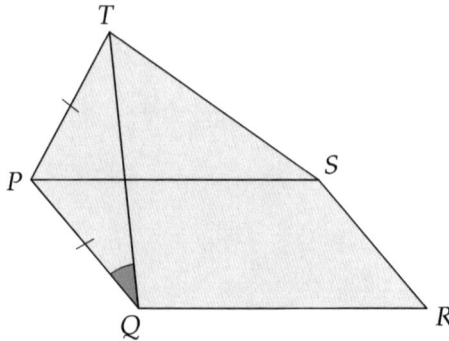

In the diagram above, the marked angle is equal to 34°.

ANSWER: ₒ96

14. In 6 minutes the minute hand turns through 36° and the hour hand turns through 3°.

ANSWER: ₒ££I

15. ☞ *The base angles of an isosceles triangle are equal.*

☞ *The angles of a triangle add up to 180°.*

The marked angle in the diagram alongside is 40°.

ANSWER: ₒ£9

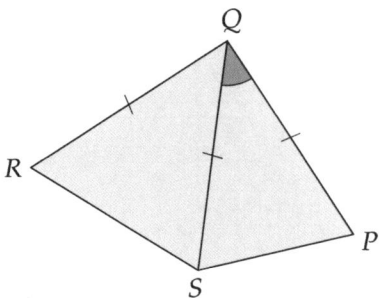

16. ☞ *Angles on a straight line add up to 180°.*

☞ *An exterior angle of a triangle is equal to the sum of the interior opposite angles.*

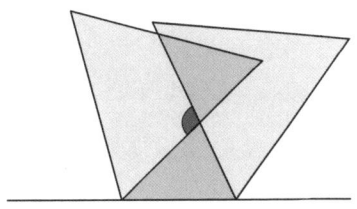

The marked angle in the diagram above is 100°.

ANSWER: ₒ0₸

17. ☞ *The vertices of a regular polygon lie on a circle.*

☞ *The base angles of an isosceles triangle are equal.*

☞ *The angles of a triangle add up to 180°.*

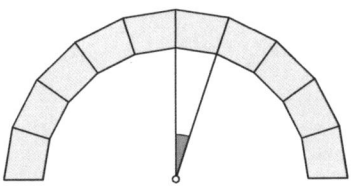

The triangle in the diagram above is isosceles, and the marked angle is 18°.

ANSWER: ₒI8

Exercise 10

1. The thickness of a single sheet is 5.4 cm ÷ 500.

   ANSWER: 0.1 mm

2. The least answer is given by $19 - 97$, and the greatest by 19^{97}.

   ANSWER: 19 + 97

3. The expression is equal to $2.017 \times (2016 - 10.16 \times 100)$.

   ANSWER: 2017

4. The number of kilograms is $60\,000\,000 \times 70$.

   ANSWER: 4 200 000 000

5. The mass of the monster in grams is $6 \times 1000 \times 1000$.

   ANSWER: 3 000 000

6. The total cost to be split was £25.50 + 2.50.

   ANSWER: £11.50

7. Mickey needs to calculate $341 \times 20 \div 1000$.

   ANSWER: 341 × 0.02

Exercise 11

1. ☞ *The exterior angles of a polygon add up to 360°.*

Answer: 120°

2. ☞ *Each interior angle of an equilateral triangle is equal to 60°.*
☞ *Angles at a point add up to 360°.*

Answer: 150°

3. ☞ *The interior angles of a polygon with n sides add up to* $(n-2) \times 180°$.

The sum of the interior angles of the final polygon is the same as the sum of the interior angles of the three separate polygons.

Answer: 1080°

4. ☞ *The exterior angle of a regular pentagon is 72°.*

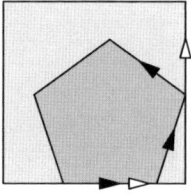

When you travel along the black arrows in the diagram alongside, you turn through $72° + 72°$. When you travel along the white arrows, you turn through 90°. Hence $90 + x = 2 \times 72$.

Answer: 54

5. ☞ *An exterior angle of a triangle is equal to the sum of the interior opposite angles.*
☞ *The exterior angles of a polygon add up to 360°.*

The sum of a pair of the marked angles is equal to an exterior angle of the central quadrilateral.

Answer: 360°

6. ☞ *The exterior angles of a polygon add up to 360°.*

Answer: 720

7. ☞ *Each angle of an equilateral triangle is equal to 60°.*

   ☞ *Each interior angle of a regular pentagon is equal to 108°.*

   ☞ *The base angles of an isosceles triangle are equal.*

   ☞ *The angles of a triangle add up to 180°.*

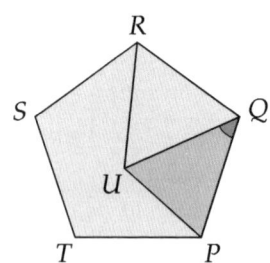

The shaded triangle in the diagram above is isosceles.

ANSWER: ₒ99

8. ☞ *Corresponding angles on parallel lines are equal.*

   ☞ *Angles at a point add up to 360°.*

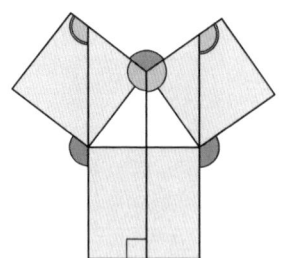

ANSWER: ₒ09ε

9. ☞ *Each interior angle of a regular pentagon is equal to 108°.*

   ☞ *The exterior angles of a polygon add up to 360°.*

The marked angle in the diagram alongside is equal to $2 \times 108° - 180°$.

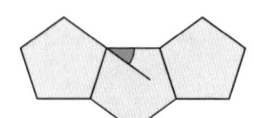

ANSWER: ∠

10. ☞ *The interior angle of a square is 90°.*

    ☞ *The interior angle of a regular hexagon is 120°.*

    ☞ *The base angles of an isosceles triangle are equal.*

    ☞ *The angles of a triangle add up to 180°.*

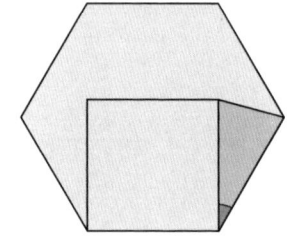

In the diagram above, the marked angle is equal to $120° - 90°$ and the shaded triangle is isosceles.

ANSWER: ₒ⊆∠

11. ☞ *A polygon with n sides has $\frac{1}{2}n(n-3)$ diagonals.*

ANSWER: The fifth statement is false.

12. ☞ *The interior angle of a regular pentagon is 108°.*

☞ *The base angles of an isosceles triangle are equal.*

☞ *The angles of a triangle add up to 180°.*

Angle *SUR* is equal to $180° - 36° - 36°$ and angles *RUP* and *PUS* are equal.

ANSWER: 126°

13. ☞ *Angles on a straight line add up to 180°.*

☞ *The exterior angles of a polygon add up to 360°.*

Each exterior angle of the polygon is a whole number of degrees, and the maximum number is obtained when each is 1°.

ANSWER: 360

Exercise 12

1. The answer is the same as 10% of the difference between one million and one thousand.

ANSWER: 006 66

2. The value is equal to $\dfrac{6}{100} \times 6 + \dfrac{8}{100} \times 8$.

ANSWER: Ɩ

3. 50% of 10 is less than 40% of 20 because 50 × 10 is less than 40 × 20.

ANSWER: 0Ɛ ⅃⁰ %0Ɛ

4. A special pack contains 50 seeds.

ANSWER: ϛƐ

5. The number of shaded cells is equal to $1 + 5 + 9$.

ANSWER: 09

6. ANSWER: 00Ɩ

7. The least number of fleas that will be eradicated by the treatment is equal to 99% of a million.

ANSWER: 000 066

8. The ratio of sheep to goats is equal to $(100 + 55) : 100$.

ANSWER: ƖƐ : 0ζ

9. S is equal to 15, 60 is equal to four-fifths of U, and $80 = \dfrac{M}{100} \times 25$.

ANSWER: 0Ɩⱨ

10. £240 is 75% of what I would have paid before the sale, and I saved 25% of this amount, so I saved one third of £240.

ANSWER: 08Ⅎ

11. £60 is four-fifths of what I would have paid before the sale.

ANSWER: £75

12. 18% of £30 is equal to 6% of £90 because 18 × 30 is equal to 6 × 90.

ANSWER: 12% of 50

13. The area of the part of the page *not* occupied by the margins is 26 cm × 36 cm.

ANSWER: 22

14. 2006% of 50 is equal to 50% of 2006.

ANSWER: 2006

15. 20% of the pupils were asked.

ANSWER: 16

16. 4 bags of Goldilocks' porridge mixture contain one fifth of a bag of wheat bran.

ANSWER: 5

Exercise 13

1. Suzy climbed just over 1560 steps in slightly less than 12 minutes.

ANSWER: 130

2. ☞ *Distance is equal to speed × time.*

☞ *5 miles is about 8 kilometres.*

The coach travels 60 × 2 miles.

ANSWER: 200

3. The birds travel roughly 15 000 km in 50 days.

ANSWER: 300 km

4. ☞ *Time is equal to distance ÷ speed.*

Turbo takes $\dfrac{12}{4}$ hours, and Harriet takes half this time.

ANSWER: 9:45 am

5. Bertie sprinted a distance of $\dfrac{5.5}{1000}$ kilometres in about $\dfrac{1}{180}$ hours.

ANSWER: 1

6. Ten gallons of honey provide enough fuel for one bee to fly about seventy million miles.

ANSWER: 70 000

7. ☞ *Time is equal to distance ÷ speed.*

Timmy takes $\dfrac{8}{20}$ hours to get to the surgery, Tammy takes $\dfrac{1.2}{4}$ hours, and Tommy takes $\dfrac{16.5}{45}$ hours.

ANSWER: Tammy, Tommy, Timmy

8. The rally lasted for 132 minutes, so that *each* of the two players hit just over 130 × 45 shots.

ANSWER: 12 000

9. Chris took nearly 84 hours to cover slightly more than 3360 km.

Answer: 40 km/h

10. ☛ *Distance is equal to speed × time.*

The cheetah kept going for $\dfrac{18}{60 \times 60}$ hours, so the distance covered was

$\dfrac{90 \times 18}{60 \times 60}$ kilometres.

Answer: 9 hours

Exercise 14

1. The four triangular regions not covered by the rectangle may be reassembled to form two squares, with sides of length 5 cm and 3 cm.

   ANSWER: $30\,cm^2$

2. ANSWER: The area of a tennis court is approximately 200 m^2.

3. The shaded region in the question has the same area as that shown in the diagram alongside; the area of this region is cut in half by the line YD.

   ANSWER: XA

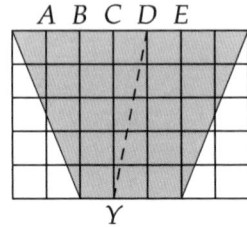

4. Each grey region may be paired with a white region with the same area, so that half of the area of the pennant is shaded grey. The area of the whole pennant is equal to three-quarters of a square with sides of length 12 cm.

   ANSWER: $54\,cm^2$

5. In the diagram alongside, PX cuts the figure into two equal parts. Hence PX cuts the figure in the question in half.

   ANSWER: The line XY crosses at X.

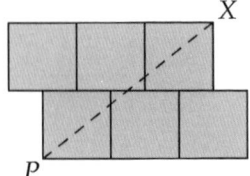

6. ☞ *The area of a triangle is equal to $\frac{1}{2} \times$ base $\times$ height.*

   ☞ *The area of a parallelogram is equal to base $\times$ height.*

   Let the whole square be 3×3. Then the area of the first shaded region is equal to $\frac{1}{2} \times 2 \times 3$.

   ANSWER:

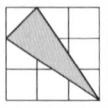

7. As shown in the diagram alongside, the figure may be cut into two pieces that can be rearranged to form a square.

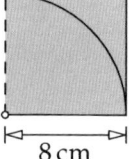

8 cm

ANSWER: 64 cm² *(shown upside down)*

8. ☞ *Triangles with the same base and equal heights have equal areas.*

The given shaded region has the same area as the shaded region in the rectangle below.

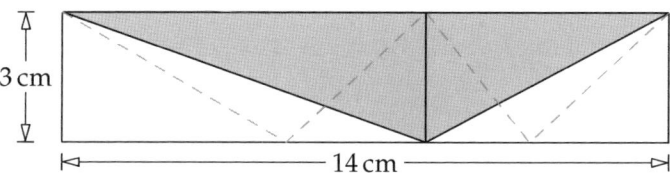

3 cm

14 cm

ANSWER: 21 cm² *(shown upside down)*

9. ☞ *Pythagoras' Theorem.*

Let the side of the square have length 2s cm; then the sides of the white triangle in the diagram alongside have lengths 1 cm, s cm and 2s cm.

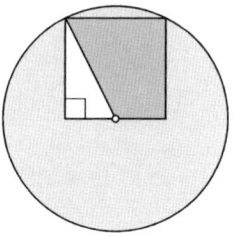

ANSWER: $\dfrac{4}{5}$ cm² *(shown upside down)*

Exercise 15

1. Each pyramid has four triangular faces. Two of these faces with an edge of the cube in common do not lie in the same plane when the solid 'star' is assembled.

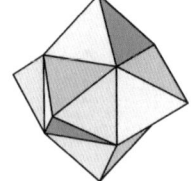

ANSWER: 24

2. Let the volume of each of the small cubes used to assemble the shapes be 1; then each triangular prism has volume $\frac{1}{2}$.

The total volume of all five shapes is $10\frac{1}{2}$. Thus only one of the shapes can be omitted to leave a volume that is a cube, and the other four shapes can actually be placed together to make a cube.

ANSWER: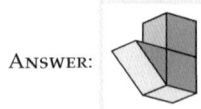

3. The tetrahedron has 6 edges and 4 vertices.

ANSWER: 24

4. As shown in the diagram alongside, it is possible to paint the faces of the octahedron using two colours, and it is clearly not possible with fewer colours.

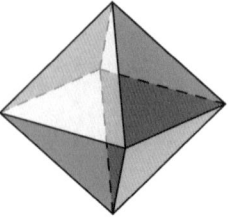

ANSWER: 2

5. The missing die has faces 1, 3 and 5 arranged *clockwise* around a vertex.

ANSWER:

6. The box has a square base and a height of 2 cm. Let the edge-length of the base be s cm; then $(s + 4)^2 = 180 + 4 \times 2^2$, so that $s + 4 = 14$.

ANSWER: 200 cm³

7. The blue faces of the small cubes form the faces of the original wooden cube.

The unpainted faces of the small cubes form both sides of slices through the original wooden cube, such as that shown in the diagram alongside.

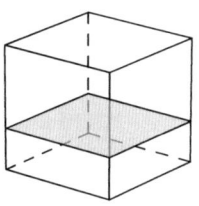

ANSWER: $\frac{1}{3}$

8. The $3 \times 2 \times 4$ cuboid has the smallest surface area.

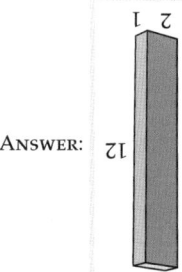

ANSWER: 12

9. A product of $4 \times 5 \times 6$ cannot be obtained.

ANSWER: 90

10. The final rhomicuboctahedron looks like that shown in the diagram alongside.

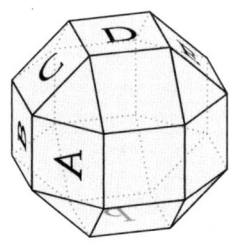

ANSWER: D

11. The ant only finishes on a grey
square when exactly one of
its moves crosses the black
vertical line shown in the
diagram alongside.

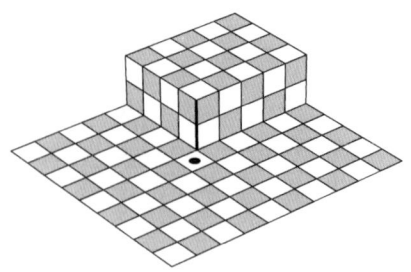

ANSWER: 4

12. The surface area of the original cube is $6 \times 10\,\text{cm} \times 10\,\text{cm}$, and the cut
only reduces the surface area by two $5\,\text{cm} \times 5\,\text{cm}$ squares.

ANSWER: $\dfrac{1}{12}$

13. The diagram alongside shows how the sculpture
appears from each of six directions.

ANSWER: 48

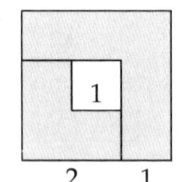

14. All the shadows are possible, as shown in the diagrams below.

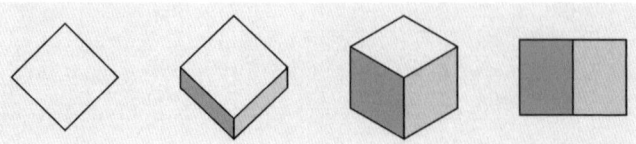

ANSWER: 4

15. There are at least three black edges.
In the diagram alongside, every face of the cube has
one black edge.

ANSWER: 3

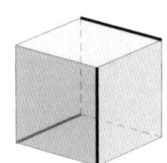

Exercise 16

1. The total number of people needed to break even over the two-week period is $14 \times 30\,000$.

ANSWER: 300 000

2. The mode is 7.
Let the missing number be x; then $7 + 7 + 5 + 7 + x = 5 \times 7$. The median of the five numbers obtained is also 7.

ANSWER: 6

3. The average cost per item in the advert is two for the price of $1\frac{1}{2}$.

ANSWER: Four for the price of three

4. $x + y + z = 3x$.

ANSWER: x

5. The total weight of all five dates was $5 \times 50\,\text{g}$.

ANSWER: 906

6. The sum of my scores on the four tests was 340.

ANSWER: 40

7. The sum of the three positive integers is 21.

ANSWER: 18

8. ☞ *For any a and b:* $a^2 - b^2 = (a - b)(a + b)$ [difference of two squares].

The sum of all 64 numbers is 64×64 and the sum of the first 36 numbers is 36×36.

ANSWER: 100

9. All the terms *after* the second term are the same.

ANSWER: 44

10. $4 \times (15 + 5 + x) = 3 \times (x + 7 + 9 + 17)$.

ANSWER: 61

11. The sum of $1.\dot{2}$ and $2.\dot{1}$ is equal to $1\frac{2}{9} + 2\frac{1}{9}$.

ANSWER: 1.6

Exercise 17

1. ☛ *An integer that is a multiple of m × n is also a multiple of m, where m and n are positive integers.*

☛ *When the highest common factor of positive integers p and q is equal to 1, an integer that is both a multiple of p and a multiple of q is also a multiple of p × q.*

2, 3 and 5 are all factors of the integer; hence 10 and 30 are also factors.

ANSWER: 30

2. The number of zeros at the end of the answer is equal to the number of factors of 5 in the multiplication.

ANSWER: 2

3. The fee for twelve players is three times the fee for four players.

ANSWER: 675

4. When $m = 2$ then $m^2 + 2 = 6$, which is not a cube.

ANSWER: 5

5. The birthday-product for someone born on 5 February 1998 is 19 980, which has 37 as a factor. Only one of the given years has 37 as a factor.

ANSWER: Queen Anne

6. Let the numbers of male and female fish be m and f respectively; then $9m + 8f = 86$.

Hence m is even, and the only possibility is $m = 6$.

ANSWER: 3 : 2

7. 12142334 is not a Langford number because the digits 3 are too close together.

ANSWER: 41312432

8. ☛ *For a positive integer d and integers m, n, where the highest common factor of both d and m and d and n is equal to 1,*

   (the remainder when m is divided by d)

   $\times$ *(the remainder when n is divided by d)*

   is equal to the remainder when m $\times$ n is divided by d.

   The remainder is equal to the remainder when $-2 \times 1 \times -1 \times -3$ is divided by 8.

   ANSWER: 2

9. No prime is abundant; and 1 is not abundant.

   ANSWER: 12

10. Seventeen houses have numbers that are divisible by 3, and eleven have numbers that are divisible by 5. But four houses have numbers that are divisible by 15.

    ANSWER: 27

11. The sum of the factors of 22 is $1 + 2 + 11 + 22$, which *is* a square.

    ANSWER: 40

12. The smallest and largest numbers are paired.

    ANSWER: 36

13. When written as a decimal, the digits of $20 \div 11$ recur in a cycle of length two.

    ANSWER: 0406

14. 9 divides $1 + d + 3 + 4 + 5 + 6$.

    ANSWER: 8

15. $y = 5$ and, from the 'thousands' and 'tens' columns, $a = 2$.

    ANSWER: 15

16. Let there be n coins in my bag; then $\frac{1}{3} \times n$, $\frac{1}{5} \times n$ and $\frac{2}{7} \times n$ are all integers.

    ANSWER: 105

Exercise 18

1. The hour hand has turned through an acute angle anticlockwise from noon.

ANSWER:

2. The two shapes can be dissected into congruent small equilateral triangles, as shown in the diagram alongside.

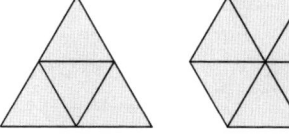

ANSWER: ɛ : ᄅ

3. The assembled shape contains twenty-one small squares, which rules out two of the smaller shapes.

The 2 × 2 square on the left of the assembled shape means that the second smaller shape was not used, and the 3 × 2 rectangle at the top excludes the fourth. But three copies of the fifth smaller shape may be used.

ANSWER: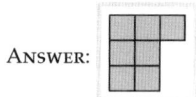

4. The first image looks like the diagram alongside.

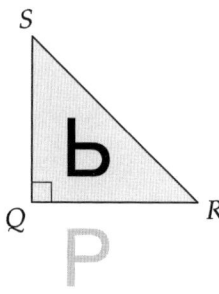

ANSWER: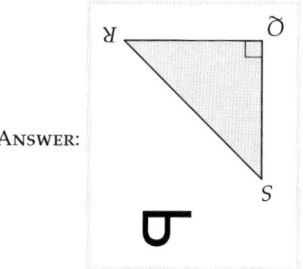

5. Triangle *XYZ* may be dissected into congruent small triangles, as shown in the diagram alongside.

ANSWER: $x\frac{6}{7}$

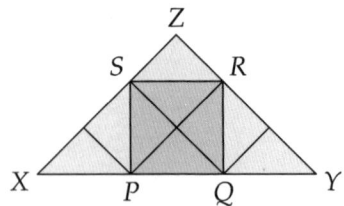

6. After the card is first turned over, to someone facing north the card appears as shown in the diagram alongside.

ANSWER:

7. The triangle may be dissected into congruent small triangles, as shown in the diagram alongside.

ANSWER: $\frac{91}{7}$

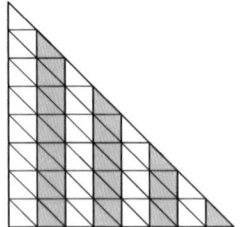

8. The square *PQRS* may be dissected into congruent small squares, as shown in the diagram alongside; for twelve of these, half the square is shaded.

ANSWER: $\frac{8}{3}$

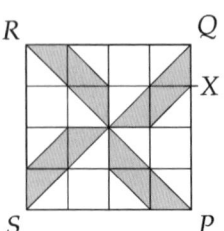

9. Pairs of the marked angles in the diagram alongside are equal, hence reflex angle *ROQ* is twice angle *NOM*.

ANSWER: 100°

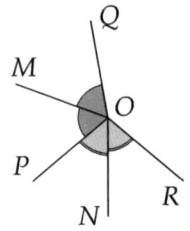

10. The difference between the areas of any two adjacent stripes is equal to the area of one of the rectangles in the diagram alongside.

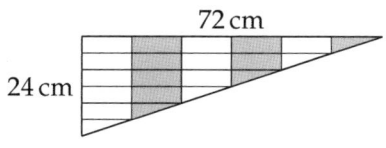

ANSWER: 48 cm² *(printed upside down)*

11. ☛ *The length of the circumference of a circle with diameter D is equal to $\pi \times D$.*

The point C traces out two arcs of a circle; the centre of one arc is the original position of B, and the centre of the other is the final position of A; each arc is one third of the circle.

ANSWER: $\frac{2}{3}\pi$ *(printed upside down)*

Exercise 19

1. Let the number be t; then $t^2(1 - 2t) = 0$ and $t > 0$.

ANSWER: $\frac{1}{2}$

2. ☞ *For any a and b:* $(a + b)^2 - (a - b)^2 = 4 \times a \times b$.

Let the greater number be a and let the other be b; then $a + b = 2$ and $a - b = 4$.

ANSWER: -3

3. Let the original number be t; then $t\left(t - \dfrac{70}{100}\right) = 0$, and $t \neq 0$.

ANSWER: 0.7

4. Suppose that Jim rolled n dice; then $2 + 3 + 5 + (n - 3) \times 1 = 2 \times 3 \times 5$.

ANSWER: 23

5. Gill's car uses $\dfrac{p}{100}$ litres of petrol for every kilometre travelled.

ANSWER: $\dfrac{pd}{100}$

6. Let the weight of the body be $b\,$kg; then $b = \left(9 + \frac{1}{3}b\right) + 9$.

ANSWER: $54\,$kg

7. $S - 1 = 5$ and $B + 1 = 3$.

ANSWER: 12

8. Let the first term be a; then the sequence is $a, 4, a + 4, a + 8, \ldots$.

ANSWER: 5

9. ☞ *For any a and b:* $(a + b)^2 - (a - b)^2 = 4 \times a \times b$.

$a + b = 7$ and $a - b = 2$ or -2.

ANSWER: $\dfrac{45}{4}$

10. Let the numbers of long- and short-sleeve shirts be ℓ and s respectively; then $8\ell + 6s = 10 \times 2\ell$.

ANSWER: 1 : 2

11. Let $T°$ Celsius be the temperature when the approximate formula gives an answer which is too large by 1; then $2T + 30 = 1 + \left(\frac{9}{5} \times T + 32\right)$.

ANSWER: 15° Celsius

12. Let the cost of a candle be c pence; then $(460 + c) + c = 610$.

ANSWER: £13.70

13. Add the three equations to get $4(x + y + z) - (x + y + z) = 30$.

ANSWER: 10

14. $(p - q) - (p + q) = -2q$, which is positive, so that $p - q$ is greater than $p + q$.

ANSWER: $p - q$

15. Let Zac's original number be z; then $\frac{1}{2}z + 8 = 2z - 8$.

ANSWER: $\frac{32}{3}$

16. There are $n - q$ people in front of Wallace, as shown in the schematic diagram below.

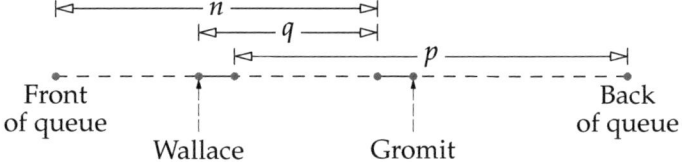

ANSWER: $n - q + p + 1$

Exercise 20

1. There are 20 illuminated dots in the letter 't'.

ANSWER: [ʇ]

2. The whole figure is divided into two quadrilaterals by each 'midline', such as the one shown in the diagram alongside.

ANSWER: ǝuịu

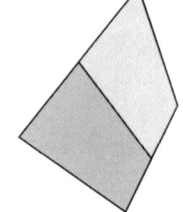

3. There are 2 × 7 + 1 cells along each side of the larger board.

ANSWER: 9ㄥꞁ

4. The numbers range from 1 + 2 to 99 + 100.

ANSWER: ㄥ6ꞁ

5. It is possible to place five T-pieces on a 5 × 5 grid.
Covering three corner cells always leaves at least two other cells uncovered, hence it is impossible to place six T-pieces on a 5 × 5 grid.

ANSWER: ǝʌg

6. When every vertex of each rectangle lies in a region of its own, as shown in the diagram alongside, then the number of regions cannot be increased.

ANSWER: ǝuịu

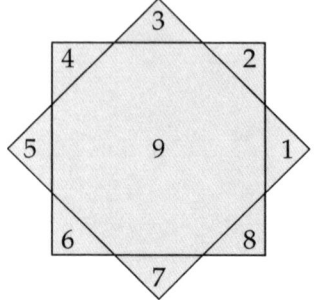

7. Having selected the first loose end, Pat may choose the second loose end in three ways.

   After Pat ties the ends together, for exactly one of these three choices will Sam hold one untied length of rope and one tied loop of rope.

   ANSWER:

8. Any quilt pattern of the type described may be represented by a cross with four arms, each grey or white, corresponding to the four edges that are sewn together (see the diagrams alongside). Each cross may have 0, 1, 2, 3, or 4 grey arms.

   ANSWER: 9

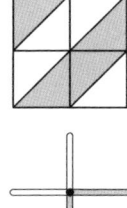

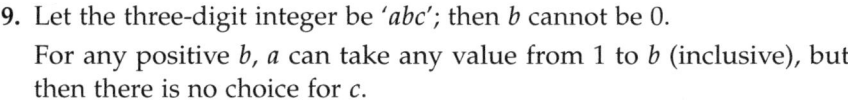

9. Let the three-digit integer be '*abc*'; then *b* cannot be 0.

   For any positive *b*, *a* can take any value from 1 to *b* (inclusive), but then there is no choice for *c*.

   ANSWER: 45

Exercise 21

1. The square formed by the centres has sides of length 2 (see the diagram alongside).

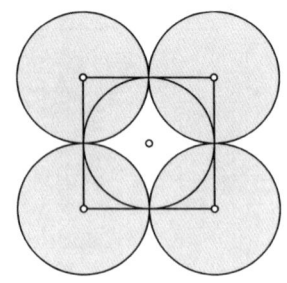

ANSWER: Ι

2. ☞ *The angle in a semicircle is 90°* [Thales' Theorem].

Draw a semicircle with centre S through *P*, *Q* and *R* (see the diagram alongside).

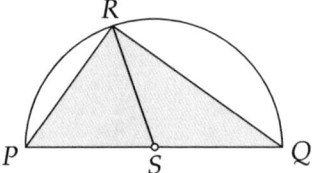

ANSWER: ₒ06

3. ☞ *The length of the circumference of a circle with diameter D is equal to* $\pi \times D$.

The required difference is equal to the total length of the 'joins', which is the length of four semicircular arcs with diameter 2 and eight straight-line segments.

ANSWER: 𝒰ƥ + 8

4. ☞ *The length of the circumference of a circle with diameter D is equal to*
$\pi \times D$.

The perimeter of the hatched shape consists of nine arcs of a circle
(see the diagram below); each arc is one sixth of the circumference of
a circle.

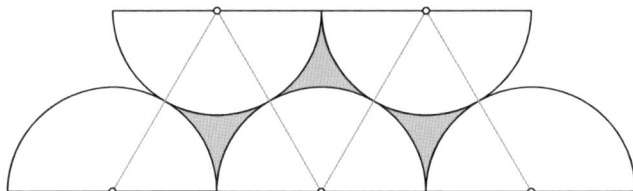

ANSWER: 9π

5. ☞ *The length of a side of a triangle is less than the sum of the lengths of the
other two sides.*

ANSWER: (!!)

6. ☞ *Pythagoras' Theorem.*

Let the third side have length ℓ cm; then either $\ell^2 = 5^2 + 6^2$ or $6^2 = \ell^2 + 5^2$.

ANSWER: two

7. ☞ *Pythagoras' Theorem.*

Let the length of the hypotenuse be h cm; then $h^2 = 125$. Also 12.5^2,
for example, is equal to $\left(\dfrac{25}{2}\right)^2$, which is $\dfrac{625}{4}$.

ANSWER: 11 cm

8. ☛ *The area of a circle with radius r is equal to $\pi \times r^2$.*

The *unshaded* (white) regions may be reassembled to form a rectangle and a semicircle, as shown in the diagram alongside; the *total* area is equal to the area of the given rectangle.

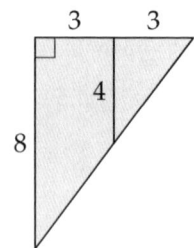

4 cm   2 cm   2 cm

ANSWER: $12 - 2\pi$

9. ☛ *The area of a circle with radius r is equal to $\pi \times r^2$.*

Let the radius of the circle be r; then the *total* area of the semicircle is equal to $\frac{1}{2} \times \pi \times (2r)^2$.

ANSWER: $\frac{1}{2}$

10. ☛ *Pythagoras' Theorem.*

The two pieces form the right-angled triangle shown in the diagram alongside.

ANSWER: 24

3   3   4   8

11. ☛ *Pythagoras' Theorem.*

The length of the diagonal of the window is 100 cm.

ANSWER: four

12. ☞ *The angles of a triangle add up to 180°.*

☞ *The base angles of an isosceles triangle are equal.*

☞ *Pythagoras' Theorem.*

☞ *The length of a side of a triangle is less than the sum of the lengths of the other two sides.*

The fourth triangle is half an equilateral triangle.

ANSWER: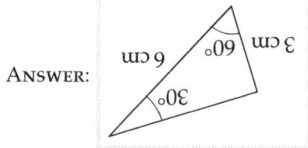

13. ☞ *The length of the circumference of a circle with diameter D is equal to* $\pi \times D$.

Let the length of the pencil be ℓ; then P moves along a semicircular arc with radius ℓ, and R moves along a semicircular arc with radius $\frac{2}{3}\ell$ followed by another semicircular arc with radius $\frac{1}{3}\ell$.

ANSWER: $1:1$

Exercise 22

1. The sum of the four given fractions is equal to $\dfrac{5}{4}$.

ANSWER: $\dfrac{5}{16}$

2. ☞ *For any a and b:* $a^2 - b^2 = (a-b)(a+b)$ [difference of two squares].

The expression is equal to $20^2 - \left(\frac{1}{2}\right)^2$.

ANSWER: $399\frac{3}{4}$

3. $\dfrac{4}{7}$ is less than $\dfrac{5}{8}$ because 4×8 is less than 7×5.

ANSWER: $\dfrac{5}{8}$

4. The second mouse ate $\frac{1}{3} \times \frac{2}{3}$ of the *original* piece of cheese.

ANSWER: $\dfrac{19}{27}$

5. ☞ *For any a and b:* *if a is less than b then a^3 is less than b^3.*

$\dfrac{5}{3}$ is greater than $\dfrac{11}{7}$ because 5×7 is greater than 3×11.

ANSWER: $\left(\dfrac{9}{5}\right)^3$

6. $b : c = (b : a) \times (a : c)$ because $\dfrac{b}{c} = \dfrac{b}{a} \times \dfrac{a}{c}$.

ANSWER: $9 : 8$

7. $\dfrac{1}{2} + \dfrac{1}{3} \times \dfrac{1}{4}$ is equal to $\dfrac{6}{12} + \dfrac{1}{12}$.

ANSWER: $\dfrac{1}{2} - \dfrac{1}{3} \times \dfrac{1}{4}$

8. $\dfrac{1}{x+2}$ is equal to $\dfrac{3.5}{1+2\times 3.5}$.

ANSWER: $\dfrac{7}{16}$

9. $\dfrac{1}{2}\times\dfrac{2}{3}$ of Noah's animals remained after the second day.

ANSWER: $\dfrac{1}{4}$

10. ☛ *The area of a triangle is equal to $\frac{1}{2}\times$ base $\times$ height.*

The height is divided by 1.25.

ANSWER: 20

11. The sequence is $\dfrac{2}{3},\ \dfrac{4}{5},\ \dfrac{11}{15},\ \dots$

ANSWER: $\dfrac{3}{4}$

12. ☛ *Time is equal to distance ÷ speed.*

Minnie's time is divided by 1.25.

ANSWER: 20

13. The area of each $8\frac{1}{2}$ inch square tin is equal to $72\frac{1}{4}$ square inches.

ANSWER: The mixture just fits (with a teeny bit of room to spare).

14. Let the original price of an item be £p; then the price is now equal to £$0.85\times\frac{1}{2}p$.

ANSWER: 57.5%

15. The *actual* percentage profit is $\dfrac{1\,500\,000-50}{50}\times 100$.

ANSWER: 3 000 000

16. The population increase for city A is equal to $\dfrac{50}{40} \times 100\%$.

ANSWER: C

17. The product is equal to

$$\frac{3}{2} \times \frac{4}{3} \times \frac{5}{4} \times \cdots \times \frac{n+1}{n},$$

which is equal to $\dfrac{n+1}{2}$.

ANSWER: The product is an integer when n is odd, and not otherwise.

18. Let n be the number of cases in 2003; then there were $1.2 \times n$ cases in 2004.

Inspector Remorse solved $0.8 \times n$ cases in 2003, and $0.6 \times 1.2 \times n$ cases in 2004.

ANSWER: The number of cases he solved went down by 10%.

19. x is less than z because

$$x = 1 - \frac{1}{111\,111},$$

whereas

$$z = 1 - \frac{3}{333\,334}$$

$$= 1 - \frac{1}{111\,111.\dot{3}}.$$

ANSWER: $y > z > x$

Exercise 23

1. ☞ *The length of the circumference of a circle with diameter D is equal to* $\pi \times D$.

   The graph is a straight line through the origin.

   ANSWER: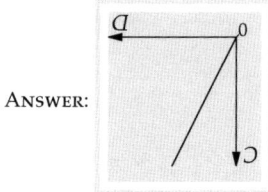

2. y increases by 2 for each increase of 1 in x between the three pairs of points: $(-3, -3)$ and $(-2, -1)$; $(-2, -1)$ and $(4, 11)$; $(4, 11)$ and $(5, 13)$. This is not the case for the pair $(-2, -1)$ and $(2, 5)$.

   ANSWER: $(2, 5)$

3. For $y = 2x + 6$, $y = 6$ when $x = 0$, and $x = -3$ when $y = 0$.

   ANSWER: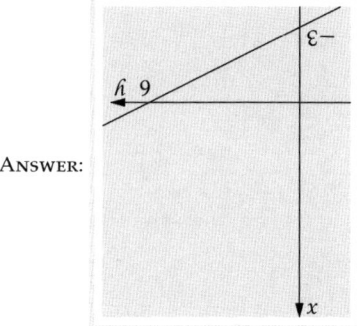

4. The graph is a straight line passing through the point $(1, 1)$.

   ANSWER: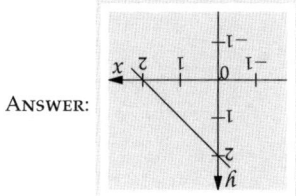

5. $(6, 2)$ is hidden by $(3, 1)$.

ANSWER: 1

6. The path of Peri's expedition is shown in the diagram alongside.

ANSWER: $(2, 0)$

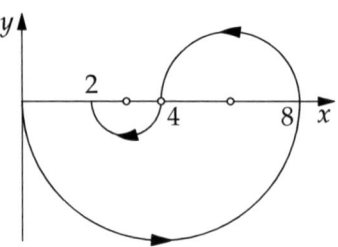

7. The triangle is the white one in the diagram below.

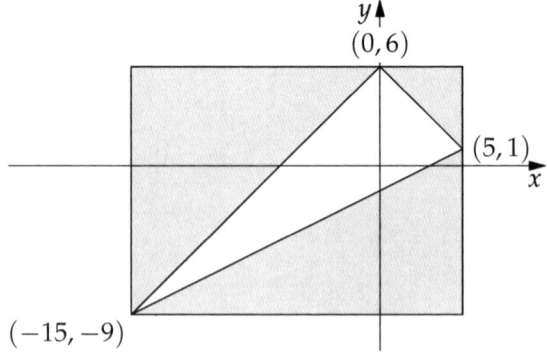

ANSWER: 75

Exercise 24

1. The displays on the clocks will next agree when one clock has gone forward 16 hours and the other has gone backward 8 hours.

   ANSWER: 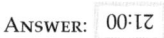 21:00

2. $8 \times 7 \times 6 \times 5 \times 4 \times 3 \times 2 \times 1$ minutes is equal to $8 \times 7 \times 6 \times 2 \times 1$ hours, which is equal to $7 \times 2 \times 2 \times 1$ days.

   ANSWER: 4

3. The watches will next agree when one has gained 6 hours and the other has lost 6 hours.

   ANSWER:  12 noon

4. The two clocks show the same time for 40 seconds in every minute.

   ANSWER: $\frac{2}{3}$

5. When the first two digits of the year are 20, any palindromic date occurs in the month of February, so that the next palindromic dates occur when the first two digits of the year are 21.

   ANSWER: 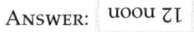 December

6. 1800 may be factorised as 12×150, and 150 is between 12×12 and 13×12.

   ANSWER: 12

7. ☞ *A year with 365 days advances the days of the week for the following year by one, and a leap year advances them by two.*

   ANSWER: 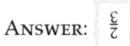 2013

8. ☞ *Suppose there are m ways of making one choice and, whichever first choice is made, n ways of making a second choice, then there are m × n ways of making both choices in succession* [multiplication principle].

The only digits which stay the same when reflected in the manner indicated are 0, 1, 3 and 8.

There are 2 choices for the first digit of the display, 4 for the second, 3 for the third, and 4 for the last.

ANSWER: 96

9. ☞ *Suppose there are m ways of making one choice and, whichever first choice is made, n ways of making a second choice, then there are m × n ways of making both choices in succession* [multiplication principle].

The first two digits of the display are fixed. There are 4 choices for the third digit, after which there are only 3 for the fifth. Finally there are only 6 choices for the fourth digit, and then 5 for the last.

ANSWER: 09Ɛ

Exercise 25

1. Beryl Burton's average speed was $\dfrac{1.2}{12}$ kilometres per hour more than Mike McNamara's.

   ANSWER: 100

2. ☞ *Distance is equal to speed × time.*

   Suppose that I take h hours to cycle back from the bike shop; then the distance to the shop is $12h$ miles. Hence I take $4h$ hours to walk to the shop.

   ANSWER: 4.8 mph

3. ☞ *Distance is equal to speed × time.*

   Let the time that Supergran needs to take be h hours; then

   $$6 \times (h + 1) = 10 \times (h - 1).$$

   ANSWER: 7.5 mph

4. ☞ *Time is equal to distance ÷ speed.*

   Suppose that the athlete walks at w mph; then

   $$\frac{1}{w} + \frac{1}{2w} + \frac{1}{3w} = \frac{3}{3w} + \frac{1}{6}.$$

   ANSWER: 22 min

5. ☞ *Distance is equal to speed × time.*

   Suppose that Mary's journey to Birmingham Airport took h hours; then

   $$60 \times h = 55 \times 2 + 70 \times (h - 2).$$

   ANSWER: 3 hours

6. When Max turns round Molly has walked $1\frac{1}{2}$ miles. Max and Molly met after Molly had walked $\frac{1}{2}$ mile further.

Answer: 4 miles

7. All three tracks have the same length.

Answer: All three reach the Finish at the same time.

8. ☞ *Distance is equal to speed × time.*

   ☞ *Time is equal to distance ÷ speed.*

Let the speed of the escalator be e m/s and the speed that Aimee walks up the escalator be w m/s; then $e \times 60 = w \times 90$.

The number of seconds Aimee takes to travel up the escalator if she walked up while it was working is $(e \times 60) \div (e + w)$.

Answer: 36

Exercise 26

1. $24 = 2^3 \times 3$.

ANSWER: One

2. When $m = 2$ and $n = 1$ then the values of three expressions are prime; it is not possible for the values of all four expressions to be prime.

ANSWER: Three

3. n cannot be odd; when $n = 2$ then $n^3 + 3$ is prime.

ANSWER: One

4. ☞ *For any p, q and positive a:* $a^p \times a^q = a^{p+q}$.

Any positive integer power of 3 is odd, and $3^{10} = \left(3^5\right)^2$.

ANSWER: (ii) and (iii)

5. The integers p and q are two of 2, 3 and 5. The highest common factor of $2p^2q$ and $3pq^2$ is obtained when $p = 3$ and $q = 5$.

ANSWER: 45

6. The missing numbers are 1, 3, 11, 13 and 19. There is only one way to complete the list, which starts as shown.

$$20, 3, 16, 15, 4, 19, 12, 1, 10, \ldots$$

ANSWER: 11

7. ☞ *A positive integer is a multiple of 9 when its digits add up to a multiple of 9, and not otherwise.*

2223 is a multiple of 9 and so $9 \times 2223 = 20\,007$ is a multiple of 81. Inserting 9 more zeros in 20 007 also gives a multiple of 81.

ANSWER: 20 000 000 000 007

8. The only way to place the first few numbers is in the arrangement shown (or in a reflection of this arrangement). The remaining numbers form the 'chain' in the box below (either clockwise or anticlockwise).

$$1 \quad 5 \quad 10 \quad 11 \quad 4 \quad 6$$

This chain has to go clockwise in the diagram above.

ANSWER: 5

9. Let n be the middle integer of the nine consecutive positive integers; then the largest number is $n + 4$ and the sum of the nine numbers is $9n$.

ANSWER: 148

10. ☞ *For any a and b:* $a^2 - b^2 = (a - b)(a + b)$ [difference of two squares].

$127^2 - 1 = 126 \times 128$.

ANSWER: 28

11. The smallest value of $a + b + c$ is achieved by setting a, b and c equal to twice the smallest possible different squares.

ANSWER: 28

12. ☞ *For any a and b:* $a^2 - b^2 = (a - b)(a + b)$ [difference of two squares].

The remainder when n^2 is divided by $n + 4$ is equal to the remainder when 16 is divided by $n + 4$ because

$$\frac{n^2}{n + 4} = \frac{n^2 - 16}{n + 4} + \frac{16}{n + 4}$$

$$= (n - 4) + \frac{16}{n + 4}.$$

ANSWER: 6

13. $\dfrac{n+3}{n-1}$ is an integer when $n-1$ divides exactly into 4, and not otherwise, because

$$\frac{n+3}{n-1} = \frac{n-1}{n-1} + \frac{4}{n-1}$$
$$= 1 + \frac{4}{n-1},$$

ANSWER: Six

14. One factor of n is 1 whatever the value of n. Paul can only get an answer of 5 by adding 1 and 4. But in that case, 2 is also a factor of n.

ANSWER: 5

Exercise 27

1. The edges of each small cube have a total length of 12 cm.

ANSWER: 124 cm

2. The cells on the board occupied by the block are those hatched or shaded in the diagram alongside.

ANSWER: 19

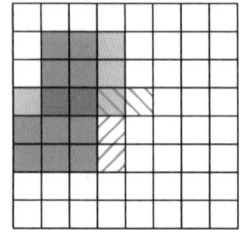

3. Each black panel is joined to five white ones, so the number of black-white joins is 12 × 5.

Each white panel is joined to three other white ones, so the number of white-white joins is $\dfrac{20 \times 3}{2}$.

ANSWER: 90

4. ☞ *The volume of a circular cylinder with radius r and height h is equal to $\pi \times r^2 \times h$.*

ANSWER: 12 : 1

5. ☞ *The area of a circle with radius r is equal to $\pi \times r^2$.*

The points that the snail could reach in one hour form the region shown shaded in the diagram alongside.

ANSWER: $1\frac{7}{8}\pi$

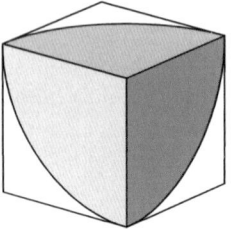

6. Consider the cuboid to be made from 60 unit cubes. After the holes are drilled, the only cubes that remain are along the edges and at the corners of the original cuboid.

ANSWER: $\dfrac{8}{15}$

7. Suppose that the lemonade is a solid shape *S*, and consider the solid shape *T* that needs to be added to *S* to fill half the can when it rests on its curved surface with *XY* horizontal.

The first two diagrams below show the cross-sections of *S* and *T* at the same distance from opposite ends of the can.

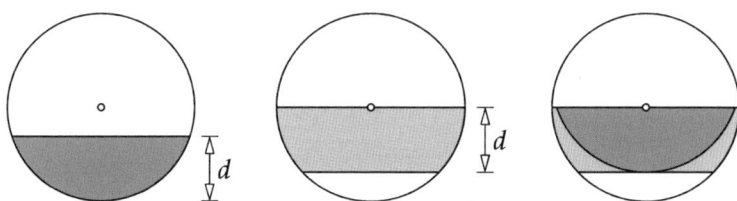

The third diagram shows that the volume of *S* is less than that of *T*.

ANSWER: just below a quarter

8. Each edge of the cube has length $\dfrac{L}{12}$, so that $6 \times \left(\dfrac{L}{12}\right)^2 = L$. Hence $L(L - 24) = 0$, and $L > 0$.

ANSWER: 8 cm³

9. The following three diagrams show the position of the die after each of the first three moves.

ANSWER: West

10. The total length of the edges of the solid and the original tetrahedron are the same.

ANSWER: 36 cm

11. The path in the diagram alongside goes along the edges of the network and passes through all 14 vertices of the network.

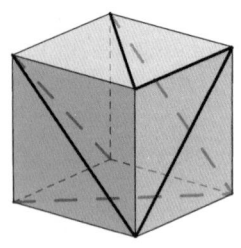

The cube has 8 corners and 6 face centres, so no path through all the vertices of the network has a shorter length than the one in the diagram.

ANSWER: 1 + 6√2

12. The dodecahedron has 20 vertices, each of which lies in three faces. Exactly nine *other* vertices lie in one of these three faces.

ANSWER: 100

Exercise 28

1. Let Pat have £p, Quentin £q, Robin £r and Sam £s; then

$$p + q + r + s = 150$$
$$p + q = 55$$
$$\text{and} \quad p + r = 65.$$

But $s - p = (p + q + r + s) - (p + q) - (p + r)$.

ANSWER: £30

2. Let a box weigh b kg, a plate p kg and a cup c kg; then

$$b + 20p + 30c = 4.8$$
$$\text{and} \quad b + 40p + 50c = 8.4.$$

It follows that $20p + 20c = 3.6$.

ANSWER: 3 kg

3. Suppose that there are b boys in Group I and g girls in Group II; then $b + (30 - g) = 37$.

ANSWER: 7

4. Since $7r = 4c$ it follows that r is less than c.

ANSWER: $r < c < s$

5. The given expression is equivalent to

$$\left(\frac{xz}{y}\right) \div \left(\frac{x}{yz}\right).$$

ANSWER: z^2

6. ☞ *For any a and b: $a^2 - b^2 = (a - b)(a + b)$ [difference of two squares].*

   The given expression is equal to

   $$(1 + x + y - 1 + x + y)(1 + x + y + 1 - x - y).$$

   ANSWER: $4(x + y)$

7. Because two woggles are the same as five waggles, and one wuggle is the same as six waggles, 2 woggles has a smaller value than 1 wuggle.

   ANSWER: 3 waggles

8. Let the length of the shorter sides of the rectangle be $4k$ cm; then the length of the longer sides of the rectangle is $5k$ cm, and $4k \times 5k = 125$.

   ANSWER: 45 cm

9. Inspector Remorse expects to take $\frac{1}{2}h$ hours to solve a bank robbery, and $\frac{1}{6}h$ hours to solve a car theft.

   ANSWER: 5h

10. Let the length of a short side of one of the rectangles be s and let the length of a long side be ℓ; then the two squares have sides of length $\ell - s$ and $\ell + s$. Hence $2(\ell - s) = \ell + s$.

    ANSWER: 1 : 3

11. $a + b + c = b - a$, so that $c = -2a$. Hence $b = -a$, $c = 2b$, $d = \frac{1}{2}c$ and $a = -d$.

    ANSWER: $\dfrac{1}{2}$

12. The maximum possible score on the first q questions is $q \times m$, and the maximum possible score on the remaining questions is $(N - q) \times (m + 2)$

    ANSWER: $(m + 2)N - 2q$

13. Let the number of chairs be c; then the number of tables is $\frac{1}{4}c$ and hence
$$3 \times \tfrac{1}{4}c + 4 \times c + 2 \times \tfrac{3}{4}c + 2 \times 3 = 206.$$

ANSWER: 32

14. It takes s men q hours to paint $r \times \dfrac{s}{p}$ square metres, and therefore it takes s men $q \times \dfrac{p}{s}$ hours to paint r square metres.

ANSWER: $\dfrac{pqr}{rs}$

15. Let the width of the traditional screen be t cm and that of a widescreen television be w cm; then the respective heights are $\dfrac{3}{4} \times t$ cm and $\dfrac{9}{16} \times w$ cm. Hence $\dfrac{3}{4} \times t \times t = \dfrac{9}{16} \times w \times w$.

ANSWER: $2 : \sqrt{3}$

Exercise 29

1. Let c be the number of cakes that Helen buys, and b the number of buns; then $39c + 23b = 512$. Hence c is at most 13.

   Answer: ~~Sixteen~~

2. The 2000-digit number consists of 400 copies of the string '12345'.

   Answer: 9000

3. Let a parsnip cost p pence and a turnip t pence; then $6p + 7t = 8p + 4t$. Hence $2p = 3t$.

   Answer: ~~Sixteen~~

4. Each zero results from multiplying one 2 and one 5 in the factorisation.

   Answer: ~~Six~~

5. ☛ *The median of an ordered list of numbers is equal to the middle number when there is one, otherwise the median is equal to the mean of the middle two numbers.*

   There are at least two 9s in the list, and the list contains one number greater than 10 and three different numbers that are at most 8.

   The list 5, 6, 7, 9, 9, 24 has all the given properties.

   Answer: 9

6. Let each smaller rectangle have length ℓ cm and breadth b cm; then $4\ell = 5b$.

   Hence $\ell = 5k$ and $b = 4k$ for some positive integer k, so that the area of the large rectangle is $9 \times 20k^2$ for some positive integer k.

   Answer: ~~1620 cm²~~

7. The calculation may be written

$$1 + (3 - 2) + (5 - 4) + \cdots + (n - (n - 1)),$$

and hence $\dfrac{n+1}{2} \times 1 = 2006$.

ANSWER: 9

8. n is not 1 and $2n + e = 8$. Hence e is even, but at most 4. However, $e = 4$ is not possible.

ANSWER: 5

9. Suppose that the Black Knight wins b bouts and loses c, and that the Red Knight wins r bouts and loses s; then

$$20b + 17c = 1 + 20r + 17s. \qquad (1)$$

Let m be the (positive) difference between b and r. Similarly, let n be the difference between c and s. Then from equation (1) the difference between $20m$ and $17n$ is 1, and 6 and 7 are the smallest values of m and n for which that is the case.

ANSWER: 13

10. Suppose the numbers of cups of coffee that the 190 customers buy are listed in increasing order, and let a and b be the 95th and 96th numbers in the list; then the median is maximised when $a + b$ is maximised.

The list of 190 numbers made up of of ninety-four 1s, one 3 and ninety-five 4s has a total of 477, and this list maximises $a + b$.

ANSWER: 3.5

11. Let a be the number of adults who went to see the film and let c be the number of children; then $100T = 770a + 420c$.

Hence $100T$ is a multiple of 7, and therefore T is a multiple of 7.

ANSWER: 91

12. Suppose that Jasmine buys p poisoned ivy plants, d deadly nightshade plants and t triffids; then

$$2p + 9d + 12t = 120 \quad \text{and}$$
$$p + d + t = 20.$$

Thus

$$7d + 10t = 80.$$

Hence 10 divides $7d$, and therefore 10 divides d.

ANSWER: ɘuO

13. Each digit of $10^{640} - 1$ is 9, and there are 640 of them.

Thus $\dfrac{10^{640} - 1}{9}$ also has 640 digits.

ANSWER: I8ZI

14. 2006 is even, so that 6^k is odd, and hence $k = 0$. Therefore $5^j + 7^\ell + 11^m = 2005$.

Whatever the (non-zero) values of j and m, the units digits of 5^j and 11^m are 5 and 1 respectively. Hence $\ell = 2$, so that $5^j + 11^m = 1956$.

ANSWER: 6

Exercise 30

1. ☞ *The area of a circle with radius r is equal to* $\pi \times r^2$.

Let the radius of the dashed circle be r cm; then $r^2 - 2^2 = 14^2 - r^2$.

ANSWER: 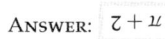 10 cm

2. ☞ *The length of the circumference of a circle with diameter D is equal to* $\pi \times D$.

The perimeter of the shaded region consists of four circular arcs, each of which is three-quarters of a circle.

ANSWER: 9

3. ☞ *The length of the circumference of a circle with diameter D is equal to* $\pi \times D$.

The perimeter of the shaded region consists of four semicircles and four quarter-circles (see the diagram alongside).

ANSWER: 3π

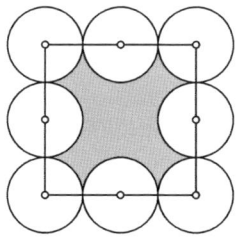

4. ☞ *The area of a circle with radius r is equal to* $\pi \times r^2$.

The shaded region may be divided into a 2×1 rectangle, a semicircle and two quarter-circles, as shown in the diagram alongside.

ANSWER: $\pi + 2$

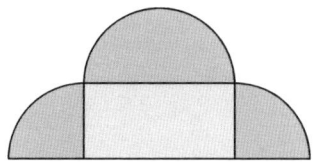

5. ☞ *The area of a circle with radius r is equal to* $\pi \times r^2$.

The shaded area is equal to $\pi \times 4^2 - \pi \times 3^2$.

ANSWER: 7π

6. ☞ *The area of a circle with radius r is equal to $\pi \times r^2$.*

Let the common radius be r and enclose the diagram in the question in a square with sides of length $2r$ (see the diagram alongside).

Then the shaded area is equal to $4r^2 - \pi r^2$.

ANSWER: $4 - \frac{\pi}{4}$

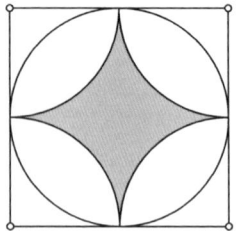

7. ☞ *The diagonals of a rhombus cross at right angles.*

☞ *Pythagoras' Theorem.*

Let P and Q be the centres of the semicircles, and let X and Y be their points of intersection (see the diagram alongside); then $PXQY$ is a rhombus, and the white triangle is right-angled.

ANSWER: 6 cm

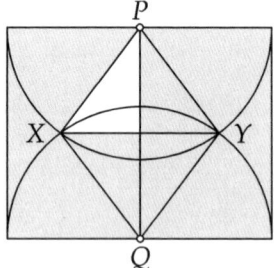

8. ☞ *The exterior angles of a polygon add up to 360°.*

☞ *The area of a circle with radius r is equal to $\pi \times r^2$.*

The total area of the shaded regions is equal to the area of five semicircles plus the area of a circle (see the diagram alongside).

ANSWER: 14π

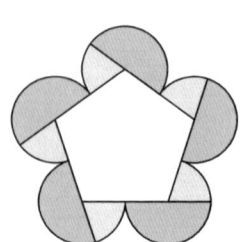

9. ☞ *In a circle, the angle at the centre is equal to twice the angle at the circumference.*

☞ *Pythagoras' Theorem.*

Let O be the centre of the circle, as shown in the diagram alongside; then angle POR is 90°.

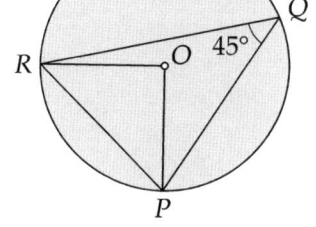

ANSWER: 4√2 cm

10. ☞ *Pythagoras' Theorem.*

☞ *The area of a circle with radius r is equal to $\pi \times r^2$.*

The two circles have the same centre, and the white triangle in the diagram alongside is right-angled. The ratio of the radii of the two circles is equal to $\sqrt{2} : 1$.

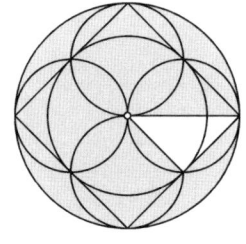

ANSWER: 2

11. ☞ *The angle in a semicircle is 90° [Thales' Theorem].*

☞ *Pythagoras' Theorem.*

☞ *The area of a circle with radius r is equal to $\pi \times r^2$.*

☞ *The area of a triangle is equal to $\frac{1}{2} \times base \times height$.*

The shaded area is equal to the area of a semicircle plus the area of one quarter of a circle minus the area of a triangle.

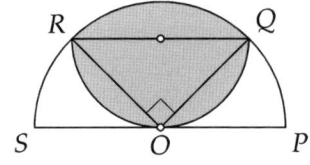

ANSWER: $2\pi - 2$

12. ☞ *Pythagoras' Theorem.*

Each arc determines a square, whose side-length is equal to the radius of an arc. These squares fit together (see the diagram alongside).

The radius of the small circle is equal to the diagonal length of one of the squares, which is 2 cm, minus the radius of an arc.

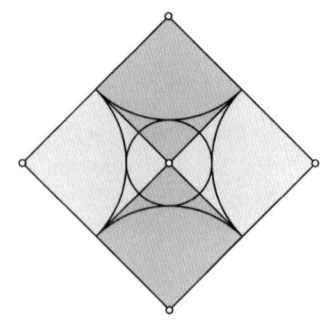

ANSWER: $(2 - \sqrt{2})\,\text{cm}$

13. ☞ *The length of the circumference of a circle with diameter D is equal to π × D.*

☞ *The area of a sector of a circle is equal to $\frac{1}{2} \times r \times s$, where r is the radius and s is the arc length.*

Let the radius of the disc be *r*, as shown in the diagram alongside, where the removed sector is unshaded; then the *arc length* of the removed sector is equal to $\pi \times 2r - 2r$.

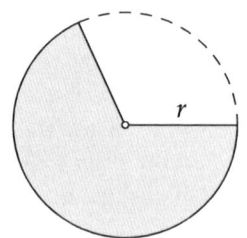

ANSWER: $\dfrac{\pi}{\pi - 1}$

14. ☞ *Pythagoras' Theorem.*

Let the radius of the circle be *r* m.

The shaded triangle shown in the diagram alongside is right-angled.

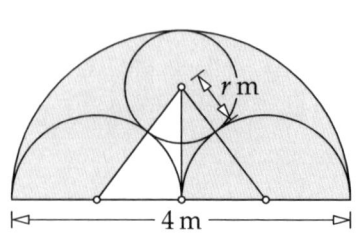

ANSWER: $\frac{2}{3}\,\text{m}$

15. ☞ *The area of a circle with radius r is equal to $\pi \times r^2$.*

☞ *The area of a square is equal to half the square of the length of a diagonal.*

The shaded region in the question consists of four *lunes* attached to the sides of the square *PQRS*; each lune is a semicircle minus a segment (see the diagram alongside).

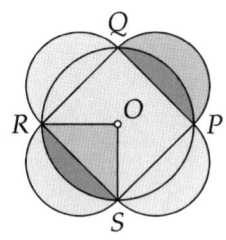

The area of each semicircle is equal to a quarter of the area of the circle, so that the area of each lune is equal to a quarter of the area of the square.

ANSWER: 2

16. ☞ *The area of a circle with radius r is equal to $\pi \times r^2$.*

The area of the shape given in the question is equal to the area of the shape shown in the diagram alongside, which is equal to the area of the triangle plus the area of a complete circle of radius $\frac{1}{2}$.

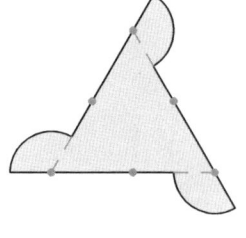

ANSWER: $1\frac{1}{4}\pi$

17. ☞ *The area of a circle with radius r is equal to $\pi \times r^2$.*

☞ *The area of a square is equal to half the square of the length of a diagonal.*

The area of segment *RSU* is twice the area of the segment *RPT* (see the diagram alongside).

Hence the given shaded area is equal to that of the square *PRQS*.

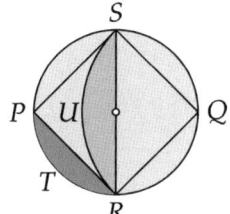

ANSWER: 2 cm²

Exercise 31

1. The number on the right of the *n*th row is n^2.
The next square after 400 is 441.

ANSWER: 441

2. The sequence is

$$6, 3, 14, 7, 34, 17, 84, 42, 21, 104, 52, 26, 13, 64, 32, 16, 8, 4, 2, 1, \ldots$$

and the cycle 4, 2, 1 now repeats.

ANSWER: 13 and 16 only

3. The total amount of platinum ever produced has a volume of about
$\dfrac{50 \times 11}{21.45}$ m^3.

ANSWER: a house

4. The numbers represented by *P* and *Q* have a product that is positive and less than the number represented by *P*.

ANSWER: B

5. 0.2008 is less than 0.2008̇.

ANSWER: 0.2008

6. ☞ *For any non-zero a:* $a^1 = a$.
☞ *For any p, q and positive a:* $a^p \times a^q = a^{p+q}$.
$4^x + 4^x + 4^x + 4^x = 4^{x+1}$.

ANSWER: 15

7. ☞ *For any p, q and positive a:* $\left(a^p\right)^q = a^{pq}$.
$4^m = \left(8^m\right)^{\frac{2}{3}}$.

ANSWER: 6

8. ☛ *Pythagoras' Theorem.*

The square folds into the trapezium in the manner shown in the diagram alongside.

ANSWER: $2 + 3^\wedge/2$

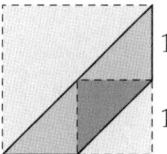

9. Because x is positive and less than 1, y is greater than $x \times y$ for any positive y.

ANSWER: $x^2 + x$

10. The pencil draws an equilateral triangle inside the wire, but draws an arc of a circle at each corner outside the wire.

ANSWER:

11. ☛ *For any p, q and positive a:* $\left(a^p\right)^q = a^{pq}$.

$5^{pq} = 12$.

ANSWER: 2

12. Nothing is gained by transferring any of the mixture.

ANSWER: It is not possible.

13. Equilateral triangles with sides of length 10, 11 and 13 may be appended to the hexagon (see the diagram alongside).

ANSWER: 27

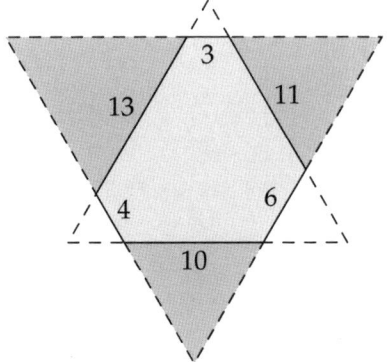

14. The only posssible rectangles made up of 32 squares have dimensions 1×32, 2×16 and 4×8.

None of these rectangles can contain four copies of the fourth shape, but it is possible to fit four copies of each of the other shapes together to make a rectangle.

ANSWER: 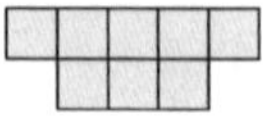

15. ☛ *Sides opposite equal angles of a triangle are equal.*

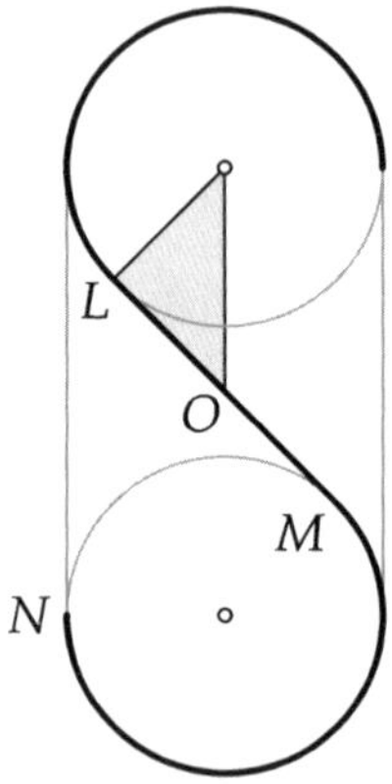

Let the midpoint of LM be O; then the shaded triangle in the diagram alongside is right-angled. One of the other angles of this triangle is 45°.

ANSWER: 2

16. The numbers leading diagonally south-east from 1 are the odd squares. The next two odd squares above 2012 are 2025 and 2209.

ANSWER: 2195

17. ☛ *For positive x less than 1: x^2 is less than x.*

☛ *For positive x less than 1: $\sqrt{x}$ is greater than x.*

☛ *For x greater than 1: x^2 is greater than x.*

☛ *For x greater than 1: $\sqrt{x}$ is less than x, but greater than 1.*

p is greater than q, so that $\frac{p}{q}$ is greater than 1.

ANSWER: $\frac{q^2}{p^2}$

18. ☞ *The area of a circle with radius r is equal to $\pi \times r^2$.*

☞ *The area of a triangle is equal to $\frac{1}{2} \times base \times height$.*

Let the radius of the semicircle be r and the height of the triangle be h (see the diagram alongside); then $\tan x^\circ = \dfrac{h}{r}$.

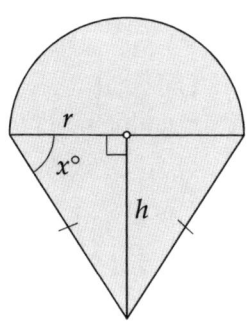

ANSWER: $\frac{1}{2}r$

Exercise 32

1. The shaded triangle in the diagram alongside is equilateral and has sides of length 12 cm.

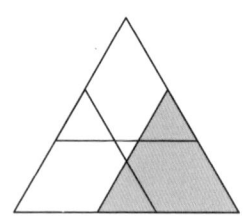

ANSWER: ⅄uɔ 6Ɩ

2. ☞ *The angle in a semicircle is 90°* [Thales' Theorem].

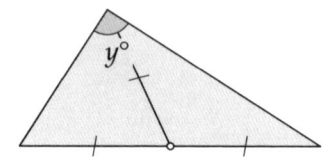

Place the two lower triangles in the given diagram side by side, as shown in the diagram alongside; then $y = x$.

ANSWER: 06

3. ☞ *Each interior angle of a regular pentagon is equal to 108°.*

☞ *Each interior angle of a regular hexagon is equal to 120°.*

☞ *The interior angles of a polygon with n sides add up to $(n-2) \times 180°$.*

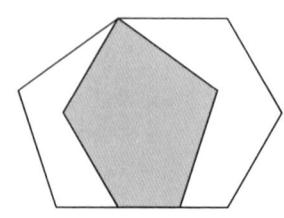

The region of overlap of the original pentagon and hexagon is shown shaded in the diagram alongside. Four of the interior angles of the shaded pentagon are 108°, 108°, 120° and 120°.

ANSWER: ⅋8

4. ☞ *Each exterior angle of a regular polygon with n sides is equal to $360° \div n$.*

☞ *The base angles of an isosceles triangle are equal.*

Triangle TUP is isosceles.

ANSWER: 9 : ⅂ : Ɛ

5. ☞ *The angles in a triangle add up to 180°.*

Angle QRP is equal to $180° - \alpha° - \beta°$ and angle QRN is equal to $90° - \beta°$.

ANSWER: $\frac{1}{2}(\alpha - \beta)°$

6. ☞ *The vertices of a regular polygon lie on a circle.*

☞ *In a circle, the angle at the centre is equal to twice the angle at the circumference.*

The marked angle is equal to $\dfrac{5}{12} \times 360°$.

ANSWER: $75°$

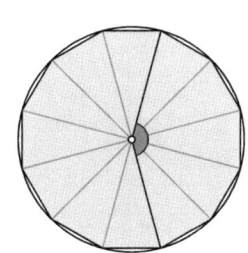

7. ☞ *Alternate angles on parallel lines are equal.*

☞ *Sides opposite equal angles of a triangle are equal.*

ANSWER: $NM = NW$

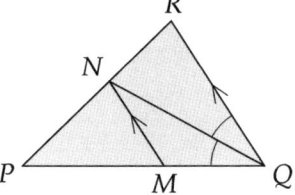

8. ☞ *The base angles of an isosceles triangle are equal.*

☞ *The angles in a triangle add up to 180°.*

☞ *Sides opposite equal angles of a triangle are equal.*

Triangle PTS is half an equilateral triangle, so that $TS = 1$. Hence triangle STR is isosceles.

Triangles TQR and PTR are also isosceles. Therefore triangle PQT is isosceles.

ANSWER: $75°$

9. ☞ *The interior angles of a polygon with n sides add up to* $(n - 2) \times 180°$.

All five of the given properties are possessed by the heptagon in the diagram alongside.

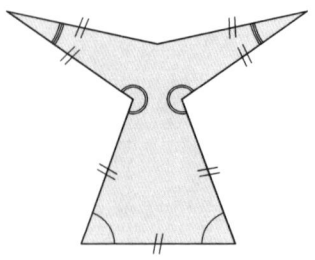

ANSWER: Five

10. ☞ *The vertices of a regular polygon lie on a circle.*

☞ *In a circle, the angle at the centre is equal to twice the angle at the circumference.*

The marked angle is equal to $360° \div 9$.

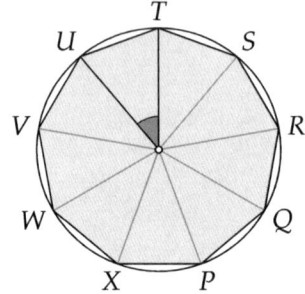

ANSWER: 20°

11. ☞ *Each interior angle of a regular hexagon is 120°.*

☞ *The diagonals of a regular hexagon bisect the interior angles.*

Create several parallelograms and equilateral triangles by drawing a diagonal of the regular hexagon, as shown in the diagram alongside. Then

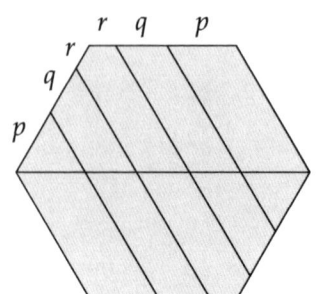

$$3p = 2q + p + (p + q)$$
$$\text{and} \quad 3p = 4r + 2(p + q).$$

ANSWER: 12 : 4 : 1

Exercise 33

1. ☞ *The base angles of an isosceles triangle are equal.*

☞ *When two angles of one triangle are equal to two angles of a second triangle the triangles are similar* [AA test for similar triangles].

The triangles PQR and RPS are similar, so that $\dfrac{SR}{6\,\text{cm}} = \dfrac{6\,\text{cm}}{9\,\text{cm}}$.

ANSWER: 4 cm

2. ☞ *An exterior angle of a triangle is equal to the sum of the interior opposite angles.*

☞ *The angles of a triangle add up to 180°.*

☞ *Pythagoras' Theorem.*

ANSWER: The first triangle is right-angled; so is the third; the other two are not.

3. ☞ *Pythagoras' Theorem.*

☞ *The area of a square is equal to half the square of the length of a diagonal.*

Let the length of the hypotenuse be h cm; then $2h^2 = 72$.

ANSWER: 9 cm²

4. ☞ *The angles of a triangle add up to 180°.*

☞ *Sides opposite equal angles of a triangle are equal.*

Angle SRP is equal to 30° and triangle PQS is half an equilateral triangle.

ANSWER: 1 : 2

5. ☞ *Pythagoras' Theorem.*

The diagram below shows the five places.

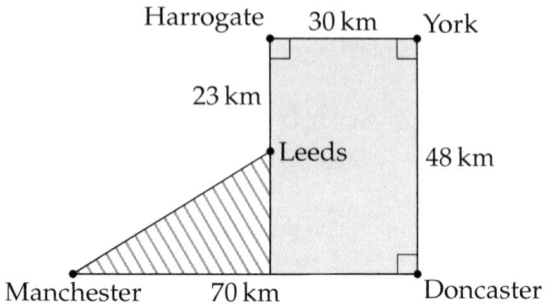

The hatched triangle is right-angled.

ANSWER: 47 km

6. ☞ *The area of a triangle is equal to $\frac{1}{2} \times base \times height$.*

The diagram below shows a 'base' of length 8 cm and a dashed line at an appropriate height.

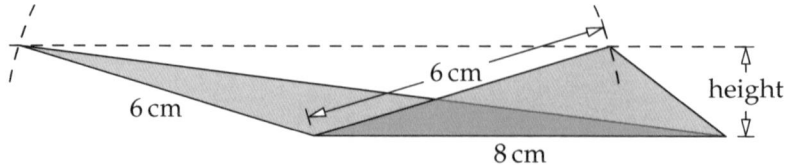

There are two points on the dashed line that are 6 cm from the left-hand end of the base. Drawing the dashed line below the base, or using the right-hand end of the base, does not change the dimensions of the triangles.

ANSWER: 2

7. ☞ *Pythagoras' Theorem.*

☞ *The area of a triangle is equal to $\frac{1}{2} \times base \times height$.*

ANSWER: 30

8. ☞ *Pythagoras' Theorem.*

Let F be the point on QR extended where $\angle QFS = 90°$, as shown in the diagram alongside.
Triangle RFS is half an equilateral triangle.

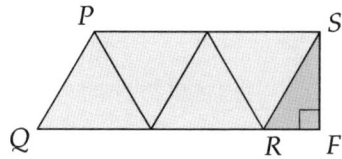

ANSWER: √2

9. ☞ *Pythagoras' Theorem.*

Join the centres, and join P and Q to the centre of the corresponding circle.

Divide the resulting trapezium into a rectangle and a right-angled triangle, as shown in the diagram alongside.

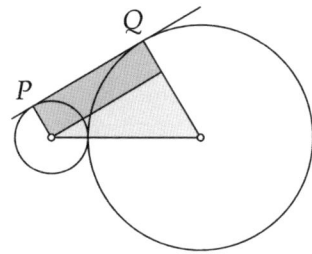

ANSWER: 4 cm

10. ☞ *Pythagoras' Theorem.*

Let the sides of the upper square have length $2x$, and those of each lower square have length y; then the areas of the upper and lower shaded regions given are $4x^2$ and $2y^2$.

Let the common radius be r. Each of the two hatched triangles in the diagrams alongside is right-angled.

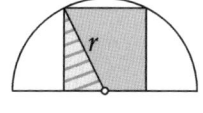

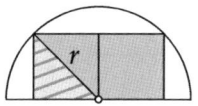

ANSWER: 4 : 5

11. ☞ *The area of a circle with radius r is equal to* $\pi \times r^2$.

☞ *Pythagoras' Theorem.*

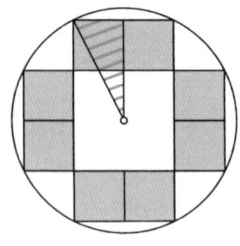

The radius of the circle is 1.

Let x be the side-length of each square. The hatched triangle in the diagram alongside is right-angled, so that $x^2 + (2x)^2 = 1^2$.

ANSWER: $1\frac{2}{5}$

12. ☞ *When two angles of one triangle are equal to two angles of a second triangle the triangles are similar* [AA test for similar triangles].

Triangles PQR and PRS are similar.

Hence $\dfrac{PR}{2\frac{1}{3}} = \dfrac{6\frac{6}{7}}{PR}$, so that $PR^2 = 6\frac{6}{7} \times 2\frac{1}{3}$.

ANSWER: 4

13. ☞ *The area of a square is equal to half the square of the length of a diagonal.*

SQ is equal to $\sqrt{12}$, and triangle SRQ is half an equilateral triangle.

ANSWER: 2

14. ☞ *The area of a square is equal to half the square of the length of a diagonal.*

☞ *Pythagoras' Theorem.*

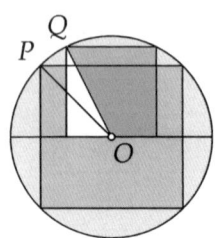

Let O be the centre of the circle, and let P and Q be the points shown in the diagram alongside; then OP and OQ are equal.

The white triangle is right-angled.

ANSWER: $2 : 5$

15. ☞ *The lengths of the diagonals of a square are equal.*

☞ *Pythagoras' Theorem.*

The two small triangles at the ends of the plank form a square with diagonal length 1 (see the diagram alongside), so that the length of the diagonal of the frame is equal to $x + 1$.

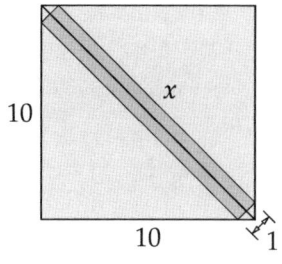

ANSWER: 1 - 2/^01

16. ☞ *Pythagoras' Theorem.*

The left-hand diagram below shows a hatched triangle in the octagon. The right-hand diagram shows this triangle divided into two right-angled triangles.

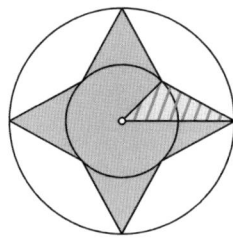 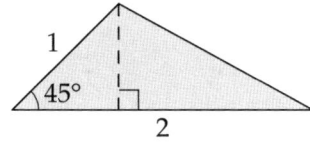

ANSWER: 2/^2 - 5/^8

17. ☛ *When two angles of one triangle are equal to two angles of a second triangle the triangles are similar* [AA test for similar triangles].

☛ *Pythagoras' Theorem.*

Let p be the length shown in the left-hand diagram below; then $p + 40 = (50 - p) + 30$, so that $p = 20$.

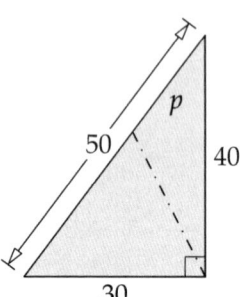 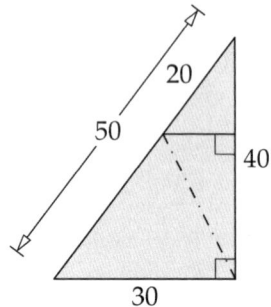

The perpendicular in the right-hand diagram creates two similar triangles.

ANSWER: ⅁/ᄼⅭⅠ

18. ☛ *Pythagoras' Theorem.*

☛ *The base angles of an isoscles triangle are equal.*

☛ *The angles in a triangle add up to 180°.*

☛ *Angles on a straight line add up to 180°.*

☛ *Sides opposite equal angles of a triangle are equal.*

The length of the complete side folded down is $\sqrt{2}$ (see the diagram alongside).

The triangle on the bottom right is a right-angled isosceles triangle.

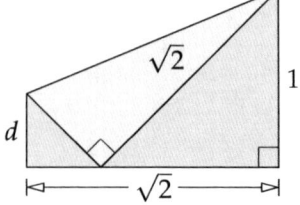

ANSWER: Ⅰ − ᄅ/ᄼ

Exercise 34

1. ☞ *The area of a circle with radius r is equal to* $\pi \times r^2$.

The area of the region in the question is one quarter of the area of the shaded region in the diagram alongside.

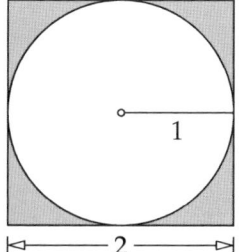

ANSWER: $\mathcal{U}\frac{t}{l} - l$

2. ☞ *The area of a triangle is equal to* $\frac{1}{2} \times$ *base* $\times$ *height.*

The total shaded area is $\frac{1}{2} \times 6 \times 8 + \frac{1}{2} \times 6 \times x$.

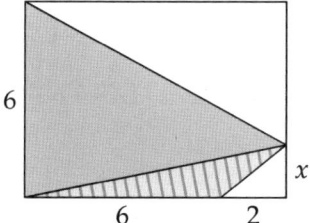

ANSWER: $\not{v}$

3. ☞ *When two angles of one triangle are equal to two angles of a second triangle the triangles are similar* [AA test for similar triangles].

Each white triangle is similar to triangle PQR. Thus the areas of the white rectangles are $\dfrac{1}{81}$ and $\dfrac{64}{81}$ of the area of $PQRS$.

ANSWER: $\dfrac{18}{9l}$

4. ☞ *Pythagoras' Theorem.*

☞ *The area of a quadrilateral with perpendicular diagonals is equal to half the product of the lengths of the diagonals.*

The lengths of the diagonals of the shaded quadrilateral are $\sqrt{2}(11 - 1 - 3)$ and $\sqrt{2}(11 - 2 - 4)$.

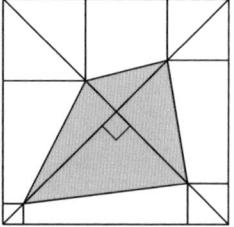

ANSWER: $35 \, cm^2$

5. ☛ *The area of a triangle is equal to* $\frac{1}{2} \times base \times height$.

The total shaded area is equal to $a \times b + \frac{1}{2} \times b \times (c - a)$.

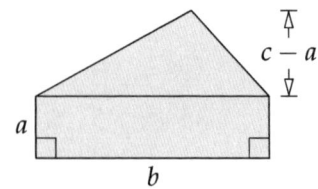

ANSWER: $\frac{1}{2}b(v+\mathfrak{d})$

6. ☛ *Pythagoras' Theorem.*

    ☛ *The area of a triangle is equal to* $\frac{1}{2} \times base \times height$.

Let the length of the unknown side of the quadrilateral be d cm; then $7^2 + 9^2 3^2 + d^2$.

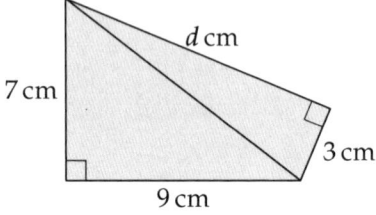

ANSWER: 48

7. Let the area of each rectangle be R, and let the area of the overlapping region be S; then the total shaded area is $2R - S$.

ANSWER: $\frac{6}{7}$

8. Let the overlapping squares have sides of length s and $2s$ (see the diagram alongside); then $s + 2s - 1 = 14$.

The total area of the shaded regions is $196 - (2s)^2 - s^2 + 1$.

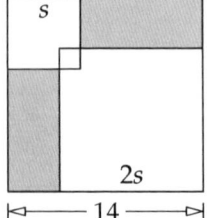

ANSWER: 77

9. ☛ *Pythagoras' Theorem.*

Let s be the length of the sides of the square $RSTU$; then s is greater than 8.

The white triangle in the diagram alongside is right-angled.

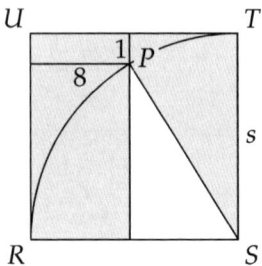

ANSWER: 13

10. ☞ *The ratio of the areas of triangles with the same height is equal to the ratio of their bases.*

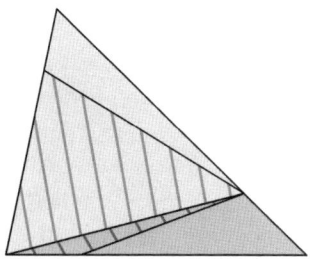

A diagonal of the given quadrilateral divides the whole triangle into two triangles (see the diagram alongside), whose areas are $\frac{1}{4}$ and $\frac{3}{4}$ of the whole area.

ANSWER: $\frac{8}{5}$

11. ☞ *When two angles of one triangle are equal to two angles of a second triangle the triangles are similar* [AA test for similar triangles].

☞ *The ratio of the areas of triangles with the same height is equal to the ratio of their bases.*

Let X be the point where the regions P and Q meet (see the diagram alongside); then X divides the diagonal of the square in the ratio 2 : 1 because the two white triangles are similar.

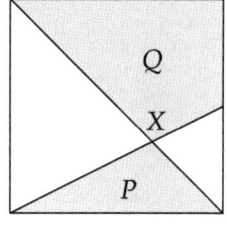

The area of the large white triangle is equal to $\frac{1}{3}$ of the area of the square, and area P is equal to $\frac{1}{6}$ of the area of the square.

ANSWER: 2 : 5

12. ☞ *Pythagoras' Theorem.*

The two overlapping squares and the regular octagon are arranged as shown in the diagram alongside. There are eight congruent right-angled isosceles triangles in the diagram.

Let the length of the legs of each of these triangles be ℓ, as shown; then $\ell + \sqrt{2}\ell + \ell = 1 + \sqrt{2}$.

The area of four of the triangles is $2\ell^2$.

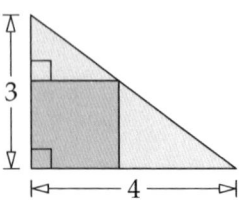

ANSWER: $2 + 2\sqrt{2}$

13. ☞ *The area of a circle with radius r is equal to $\pi \times r^2$.*

☞ *The area of a triangle is equal to $\frac{1}{2} \times base \times height$.*

Let the radius of the large circles be 2; then the areas of the five shaded regions are 2π, $3\sqrt{3}$, $16 - 4\pi$, 4 and $4\pi - 8$.

ANSWER: The first shaded region has the largest area.

14. ☞ *When two angles of one triangle are equal to two angles of a second triangle the triangles are similar* [AA test for similar triangles].

The triangle at the top of the diagram alongside is similar to the whole triangle.

Let the side-length of the square be s; then
$$\frac{3 - s}{3} = \frac{s}{4}.$$

ANSWER: $\frac{24}{49}$

Exercise 35

1. ☞ *Suppose there are m ways of making one choice and, whichever first choice is made, n ways of making a second choice, then there are m × n ways of making both choices in succession* [multiplication principle].

   (i) Three different straight lines parallel to the y-axis meet four different straight lines parallel to $y = x$ in exactly 3×4 different points.

  (ii) Two different straight lines parallel to the x-axis and four different straight lines parallel to $y = x$ meet in exactly 8 different points.

 (iii) Three different straight lines parallel to the y-axis meet two different straight lines parallel to the x-axis in exactly 6 different points.

It is possible to arrange for 6 of the 8 points in (ii) and all of the points in (iii) to coincide with other crossing points (see the diagram alongside). However, the total number of crossing points cannot be made any smaller.

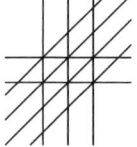

A\textsc{nswer}: ⊥Ɩ

2. ☞ *Suppose there are m ways of making one choice and, whichever first choice is made, n ways of making a second choice, then there are m × n ways of making both choices in succession* [multiplication principle].

There are 3 ways to choose the first character, 10 ways to choose a digit for the second character, and 9 ways to choose a different digit for the third character, but the final character is then determined.

A\textsc{nswer}: 0ㄥᄅ

3. The integer n lies between 6^2 and 8^2 (but is not equal to either).

A\textsc{nswer}: ㄥᄅ

4. ☞ *The number of ways of selecting 2 objects from n objects (without replacing them) is equal to* $\dfrac{n \times (n-1)}{2}$.

The two numbers *not* selected add up to a multiple of 3. One third of the number of ways of selecting two different numbers from 1 to 9 have a total that is a multiple of 3.

ANSWER: ⊥Շ

5. The routes may be paired: for every route that initially moves into the room on the right, there is a corresponding route that initially moves into the room below, and vice versa.

It is possible to enter the central room twice, but every other room can only be entered once.

The following three diagrams show different routes that initially move right.

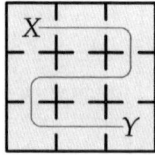 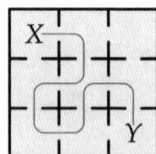 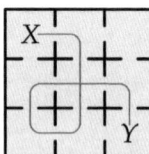

ANSWER: 9⊥

6. ☞ *The number of ways of selecting m objects from n identical objects is equal to the number of ways of selecting n − m objects.*

There are 10 lines of four points, and 4 diagonal lines of exactly three points.

ANSWER: ቱቱ

7. ☞ *The number of ways of selecting 2 objects from n objects (without replacing them) is equal to* $\dfrac{n \times (n-1)}{2}$.

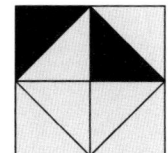

The diagram alongside has no line of symmetry; nor does any diagram obtained from it by rotation or reflection, but any other possible diagram has at least one line of symmetry.

ANSWER: $\dfrac{7}{5}$

8. After all the counters have been placed the board looks like the diagram alongside.

ANSWER: $\dfrac{23}{64}$

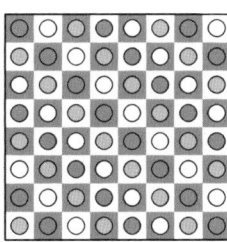

9. When each of 101 students is awarded a different mark from 0 to 100, each of the remaining students will receive the same mark as some other person.

Not awarding one of the marks from 0 to 100 initially would increase the number of pairs with the same mark. Therefore the method described above gives the smallest possible number of pairs of students who are awarded the same mark.

ANSWER: 19

10. The number of arrangements which follow UKIMC in dictionary order is $2 + 6$.

ANSWER: 11251

11. ☞ *Suppose there are m ways of making one choice and, whichever first choice is made, n ways of making a second choice, then there are m × n ways of making both choices in succession* [multiplication principle].

There are $9 \times 9 \times 8 \times 7$ four-digit integers *without* a digit repeated, and there are 9000 four-digit integers altogether.

ANSWER: ̄4464 ̄

12. There are $2^5 - 1$ possibilities that contain *exactly* two digits 3 and *d*. Hence there are 31×9 possible phone numbers with exactly two *different* digits.

ANSWER: 280

13. For any square with an even number of entries the sum of the entries is even.

The total is also even for a square with an odd number of entries provided the central entry is even, but not otherwise.

ANSWER: 36

14. 4 DOWN is 49, from 5 ACROSS and 3 ACROSS.

5 ACROSS is 19, from 2 DOWN and 1 ACROSS.

It follows that 2 DOWN is 121 and 3 ACROSS is 324.

ANSWER: Three

Exercise 36

1. The only way for Gwyn to satisfy the given requirements is to cut the L-shape into two pieces in the manner shown in the diagram alongside (or a reflection of this).

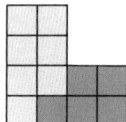

ANSWER: 2

2. From the totals for the left column and the top row, the top left number is 3, with 9 to its right and 1 below.

From the middle column and the middle row, the number in the centre is 2 (see the diagram alongside).

			Total
3	9		12
1	2		7
			13
Total 4	16	12	

ANSWER: 5

3. The maximum score is 135. Omitting a question reduces the score by 5 or 6, and doing a question incorrectly reduces the score by 5, 7 or 8.

ANSWER: 126

4. Each truth-teller has a liar on both sides, but each liar either has a liar on one side and a truth-teller on the other, or truth-tellers on both sides.

The largest number of truth-tellers occurs when every liar has a truth-teller on each side; in that case truth-tellers and liars alternate round the table.

The smallest number of truth-tellers occurs when every liar has a liar on one side and a truth-teller on the other; in that case the people round the table recur in the pattern

truth-teller, liar, liar, truth-teller, liar, liar, ...

ANSWER: 336

5. Alfred's statement cannot be true.

Bernard is not the youngest, and he is therefore telling the truth. It follows that he is not the oldest.

ANSWER: Bernard and Inigo are telling the truth.

6. The totals for the two lines containing C are equal, so that K is four more than M.

ANSWER: 13

7. There are three moves left for X.

X loses if she plays in any three of the four hatched cells in the diagram alongside.

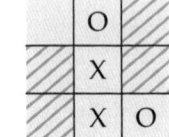

ANSWER: A

8. The first two statements cannot both be true at the same time, and the same applies to the third and fourth statements. This means that at most three statements are true at the same time.

It is possible to find a value of x for which three of the statements *are* true.

ANSWER: 3

9. The ten consecutive totals are 30 to 39 and the unused numbers are 1, 2, 8, 15 and 16. Hence only 8 can go in the bottom right corner.

ANSWER: 15

10. At most one knave can be telling the truth.

It is impossible for none to be telling the truth.

ANSWER: 1

11. With nine coins the fake coin can be identified in two weighings.

Two weighings have 3×3 possible outcomes, and hence it is not possible to identify a fake coin out of more than nine coins.

ANSWER: 6

Exercise 37

1. Each of the triangles *T* and *U* may be dissected into two 3-4-5 right-angled triangles, as shown in the diagrams below.

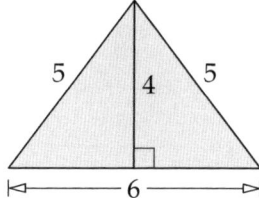 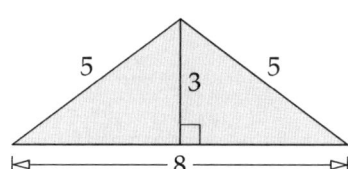

ANSWER: 1 : 1

2. The given diagram may be dissected into 30 congruent small equilateral triangles, as shown in the diagram alongside.

ANSWER: $\frac{5}{1}$

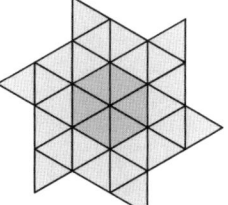

3. The octagon may be dissected into four congruent rectangles and eight congruent triangles (see the diagram alongside).

ANSWER: 4 : 1

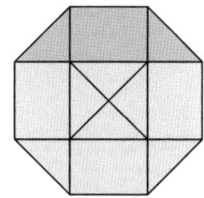

4. Rotate the shaded square through 45° about its centre, as shown in the diagram alongside.

Drawing the diagonals of the shaded square dissects the outer square into eight congruent triangles.

ANSWER: $\frac{2}{1}$

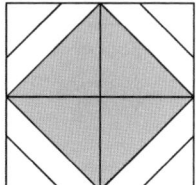

5. ☞ *The length of the circumference of a circle with diameter D is equal to π × D.*

The shape may be divided into six congruent small equilateral triangles and three 60° sectors of the circle (see the diagram alongside).

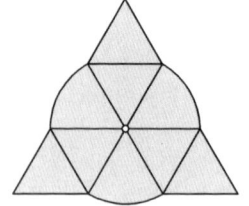

ANSWER: $\frac{u+9}{}$ (upside down: $9 + u$)

6. The large square may be dissected into 4 × 12 × 12 congruent triangles (see the diagram alongside).

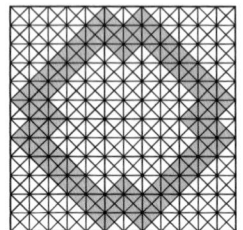

ANSWER: $\frac{\varepsilon}{\mathfrak{l}}$ (upside down: $\frac{1}{3}$)

7. The tiling may be produced by repeating a polygon P with six sides. Each copy of the polygon may be dissected into 25 congruent squares (see the diagram alongside).

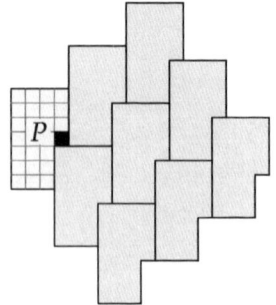

ANSWER: $\frac{\varsigma\zeta}{\mathfrak{l}}$ (upside down: $\frac{1}{25}$)

8. The octagon may be dissected into four congruent rectangles and eight congruent triangles (see the diagram alongside).

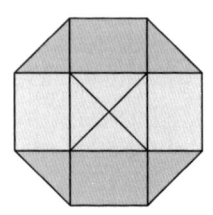

ANSWER: 0

9. The tiling may be produced by repeating a polygon *P* with eight sides. Each copy of the polygon may be dissected into 26 congruent equilateral triangles (see the diagram alongside).

ANSWER: $\dfrac{1}{13}$

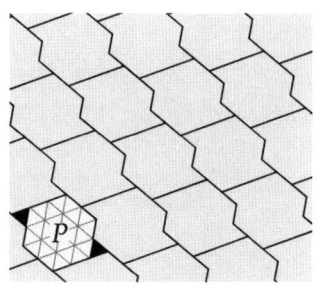

10. The tile may be divided into eight congruent triangles, each of which may be dissected into nine congruent small triangles (see the diagram alongside).

ANSWER: $\dfrac{2}{9}$

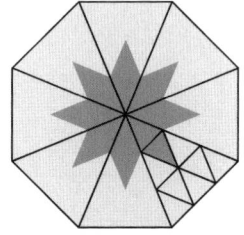

11. The area of the strip in which the rectangle lies is equal to half the area of the square.

The strip may be dissected into ten congruent triangles, as shown in the diagram alongside.

ANSWER: $\dfrac{2}{5}$

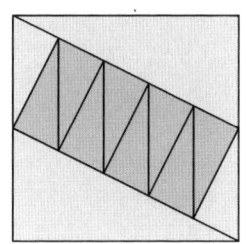

Appendix

Sources of the problems

The problems are taken from the IMC papers for the years 1997–2016. The wording of some problems has been edited, and in every case the multiple-choice options have been removed.

The following tables give the sources of all the problems in the book.

264 Intermediate Problems

Exercise 1

1.	2016 01
2.	2015 01
3.	2011 01
4.	2010 01
5.	2009 01
6.	2005 01
7.	2004 01
8.	2001 01
9.	2011 02
10.	2002 02
11.	2004 03
12.	2003 03
13.	1999 04
14.	1997 04
15.	2007 05
16.	1999 05

Exercise 2

1.	2012 01
2.	2014 02
3.	2009 02
4.	2008 02
5.	2016 03
6.	2010 03
7.	2006 03
8.	2001 06
9.	2012 07

Exercise 3

1.	2014 01
2.	2006 01
3.	2003 01
4.	2002 01
5.	1998 01
6.	2016 02
7.	2015 03
8.	2008 03
9.	2001 03
10.	2000 03
11.	2007 04
12.	1998 04
13.	2010 05
14.	2008 05
15.	1997 05
16.	2016 06
17.	2009 06
18.	1999 07

Exercise 4

1.	2013 01
2.	2000 01
3.	2012 02
4.	1997 02
5.	2011 03
6.	2002 03
7.	2016 04
8.	2014 04
9.	2012 04
10.	2001 04
11.	2002 05
12.	2011 06
13.	2007 06
14.	2003 06
15.	2000 06

Exercise 5

1.	2008 01
2.	2007 01
3.	2013 02
4.	1999 03
5.	2010 04
6.	2014 05
7.	2012 05
8.	1999 06
9.	2010 07
10.	2010 08
11.	2008 08
12.	2006 08
13.	2005 08
14.	2004 08
15.	2001 09
16.	2000 09
17.	2010 10
18.	2003 10
19.	1999 10
20.	1998 11

Exercise 6

1.	2000 02
2.	1999 02
3.	2007 03
4.	1997 03
5.	2002 04
6.	2000 04
7.	2016 05
8.	2004 06
9.	2002 06
10.	1998 06
11.	1998 08
12.	2013 09
13.	2006 09
14.	1999 09
15.	2009 11
16.	2013 12
17.	2001 13
18.	1998 13
19.	2008 15
20.	2002 15

Exercise 7

1.	1999 01
2.	1997 01
3.	2005 04
4.	2015 05
5.	2003 05
6.	2013 06
7.	2015 08
8.	2002 08
9.	1997 13
10.	2011 14
11.	2009 14
12.	2001 15

Exercise 8
1. 2006 02
2. 2004 02
3. 2003 02
4. 2005 05
5. 2000 05
6. 1998 05
7. 2006 07
8. 2012 08
9. 1998 09
10. 2014 11
11. 2010 11
12. 2002 12
13. 2016 13
14. 2002 13
15. 2016 14

Exercise 9
1. 1998 03
2. 2011 04
3. 2009 04
4. 2004 04
5. 2001 05
6. 2010 06
7. 2016 07
8. 2011 07
9. 2003 08
10. 1999 08
11. 2011 09
12. 2016 10
13. 2012 11
14. 2006 11
15. 1997 11
16. 2005 13
17. 2005 14

Exercise 10
1. 2001 07
2. 1997 07
3. 2016 08
4. 2003 09
5. 2011 10
6. 2006 10
7. 1997 10

Exercise 11
1. 2010 02
2. 2014 03
3. 2012 03
4. 2015 04
5. 2005 07
6. 2015 09
7. 2013 10
8. 2007 10
9. 2001 10
10. 2009 12
11. 1999 14
12. 2007 15
13. 2003 15

Exercise 12
1. 2005 02
2. 1998 02
3. 2009 03
4. 2013 04
5. 2008 04
6. 1997 08
7. 2012 09
8. 2009 09
9. 2008 11
10. 2000 11
11. 1999 11
12. 2012 12
13. 2007 13
14. 2006 13
15. 1997 14
16. 2005 15

Exercise 13
1. 2015 02
2. 2003 04
3. 2013 05
4. 2006 06
5. 2016 09
6. 2005 12
7. 2000 12
8. 2012 13
9. 2014 14
10. 1998 14

Exercise 14
1. 2011 05
2. 1997 06
3. 2004 07
4. 2002 07
5. 2009 08
6. 2005 09
7. 2004 12
8. 2015 13
9. 2016 15

Exercise 15
1. 2006 05
2. 2008 06
3. 2015 07
4. 2013 07
5. 2014 08
6. 2011 08
7. 2008 09
8. 2004 09
9. 1997 09
10. 2016 11
11. 2015 12
12. 2010 12
13. 2008 12
14. 2006 12
15. 2003 12

Exercise 16
1. 2001 02
2. 2006 04
3. 2005 06
4. 2007 08
5. 2000 10
6. 1998 10
7. 2015 11
8. 2008 13
9. 2015 14
10. 2012 14
11. 2006 15

Exercise 17
1. 2014 07
2. 2007 07
3. 2003 07
4. 2000 07
5. 1998 07
6. 2007 09
7. 2002 09
8. 2015 10
9. 2012 10
10. 2009 10
11. 2013 11
12. 2007 11
13. 2011 12
14. 1997 12
15. 2004 14
16. 2013 15

Exercise 18

1.	2005	03
2.	2015	06
3.	2014	06
4.	2012	06
5.	2001	08
6.	2001	11
7.	2014	12
8.	2013	13
9.	2003	14
10.	2015	15
11.	2011	15

Exercise 19

1.	2009	05
2.	2004	05
3.	2008	07
4.	2013	08
5.	2014	09
6.	2005	10
7.	2002	10
8.	2002	11
9.	2016	12
10.	2001	12
11.	1998	12
12.	2010	13
13.	2010	14
14.	2007	14
15.	2014	15
16.	1998	15

Exercise 20

1.	2007	02
2.	2013	03
3.	2004	11
4.	2014	13
5.	2003	13
6.	2000	13
7.	2008	14
8.	2001	14
9.	2000	15

Exercise 21

1.	2009	07
2.	2000	08
3.	2010	09
4.	2014	10
5.	2008	10
6.	2005	11
7.	1999	12
8.	2009	13
9.	1999	13
10.	2006	14
11.	2002	14
12.	2004	15
13.	1999	15

Exercise 22

1.	2004	10
2.	2011	11
3.	2003	11
4.	2011	13
5.	2013	14
6.	2000	14
7.	2012	15
8.	2009	15
9.	1997	15
10.	2014	16
11.	2008	16
12.	2003	16
13.	1998	16
14.	2011	17
15.	2006	17
16.	2000	20
17.	1999	21
18.	2005	22
19.	2002	25

Exercise 23

1.	2007	12
2.	2004	13
3.	2010	15
4.	2013	16
5.	2014	18
6.	2012	18
7.	2005	23

Exercise 24

1.	2010	16
2.	2014	17
3.	2007	17
4.	2001	17
5.	2003	18
6.	2000	19
7.	2002	21
8.	2008	23
9.	2002	24

Exercise 25

1.	1997	16
2.	2002	17
3.	2010	18
4.	2001	18
5.	2009	19
6.	2011	20
7.	2003	21
8.	2016	24

Exercise 26

1.	2016	16
2.	2015	16
3.	2009	16
4.	1999	16
5.	2009	18
6.	1998	18
7.	2007	19
8.	2005	19
9.	2005	20
10.	2004	20
11.	2014	22
12.	2007	23
13.	2001	23
14.	2011	24

Exercise 27

1.	2007	16
2.	2005	16
3.	2015	17
4.	2010	17
5.	2010	19
6.	2004	19
7.	2002	19
8.	1998	20
9.	2007	22
10.	2001	22
11.	2005	24
12.	1997	24

Exercise 28

1.	2006	16
2.	2015	18
3.	1999	18
4.	1997	18
5.	2013	19
6.	2008	19
7.	1999	19
8.	2012	20
9.	1999	20
10.	2004	21
11.	1997	21
12.	2004	22
13.	1999	22
14.	2003	23
15.	2003	25

Exercise 29

1.	2001	16
2.	2000	16
3.	2000	17
4.	2000	18
5.	2016	19
6.	2001	19
7.	2006	20
8.	2003	20
9.	2014	21
10.	2012	22
11.	2008	22
12.	2000	24
13.	2013	25
14.	2006	25

Exercise 30

1.	2016	17
2.	2013	17
3.	2008	17
4.	2003	17
5.	1999	17
6.	2012	19
7.	2016	20
8.	2014	20
9.	2007	20
10.	2007	21
11.	2006	21
12.	2016	23
13.	2014	23
14.	2011	23
15.	1997	23
16.	2013	24
17.	2001	25

Exercise 31

1.	2004	16
2.	2012	17
3.	2005	17
4.	2004	17
5.	2008	18
6.	2006	18
7.	2014	19
8.	2009	20
9.	2001	21
10.	2000	21
11.	2011	22
12.	2010	22
13.	2002	22
14.	1998	22
15.	2004	23
16.	2012	24
17.	2004	24
18.	2007	25

Exercise 32

1.	2012	16
2.	2002	16
3.	2006	19
4.	1998	19
5.	2008	21
6.	2003	22
7.	1997	22
8.	2012	23
9.	2002	23
10.	2000	23
11.	2005	25

Exercise 33

1.	2009	17
2.	1998	17
3.	2016	18
4.	2004	18
5.	2011	19
6.	2013	20
7.	1997	20
8.	2012	21
9.	2005	21
10.	2013	22
11.	2010	23
12.	2006	23
13.	2008	24
14.	2006	24
15.	2003	24
16.	2014	25
17.	2008	25
18.	1999	25

Exercise 34

1.	1997	17
2.	2005	18
3.	2002	18
4.	2015	19
5.	2003	19
6.	2008	20
7.	2015	21
8.	2013	21
9.	2015	24
10.	2015	25
11.	2011	25
12.	2010	25
13.	2000	25
14.	1998	25

Exercise 35

1.	1997	19
2.	2015	20
3.	2010	20
4.	2001	20
5.	2010	21
6.	2009	21
7.	2009	22
8.	2006	22
9.	2000	22
10.	2015	23
11.	2014	24
12.	2010	24
13.	2007	24
14.	1997	25

Exercise 36

1.   2011   16
2.   2007   18
3.   2002   20
4.   2016   22
5.   2013   23
6.   1999   23
7.   1998   23
8.   2009   24
9.   2001   24
10.   1999   24
11.   2009   25

Exercise 37

1.   2013   18
2.   2011   18
3.   2016   21
4.   2011   21
5.   1998   21
6.   2015   22
7.   2009   23
8.   1998   24
9.   2016   25
10.   2012   25
11.   2004   25

Index

Only terms from the problems appear in the index. Please also refer to the table of contents.